W9-ANJ-523

1118

Simple and Usable

web, mobile, and interaction design

Second Edition

Giles Colborne

Wilbraham Public Library
25 Crane Park Dr.
Wilbraham, MA 01095

New Riders | VOICES THAT MATTER™

Simple and Usable Web, Mobile, and Interaction Design, Second Edition
Giles Colborne

New Riders
www.newriders.com

Copyright © 2018 by Giles Colborne
All Rights Reserved.

New Riders is an imprint of Peachpit, an imprint of Pearson Education, Inc.

To report errors, please send a note to errata@peachpit.com

Notice of Rights
Printed in the United States of America. This publication is protected by
copyright, and permission should be obtained from the publisher prior to
any prohibited reproduction, storage in a retrieval system, or transmission in
any form or by any means, electronic, mechanical, photocopying, recording,
or otherwise. For information regarding permissions, request forms, and the
appropriate contacts within the Pearson Education Global Rights & Permissions
department, please visit www.pearsoned.com/permissions/.

Notice of Liability
The information in this book is distributed on an "As Is" basis, without
warranty. While every precaution has been taken in the preparation of this
book, neither the author nor Peachpit shall have any liability to any person
or entity with respect to any loss or damage caused or alleged to be caused
directly or indirectly by the instructions contained in this book or by the
computer software and hardware products described in it.

Trademarks
Unless otherwise indicated herein, any third-party trademarks that may appear in
this work are the property of their respective owners, and any references to third-
party trademarks, logos, or other trade dress are for demonstrative or descriptive
purposes only. Such references are not intended to imply any sponsorship,
endorsement, authorization, or promotion of Pearson Education, Inc. products by
the owners of such marks, or any relationship between the owner and Pearson
Education, Inc. or its affiliates, authors, licensees, or distributors.

Executive Editor: Nancy Davis
Development and Project Editor: Victor Gavenda
Senior Production Editor: Tracey Croom
Copy Editor: Dan Foster
Proofreader: Kim Wimpsett
Compositor: Danielle Foster
Indexer: Rebecca Plunkett
Cover and Interior Design: Mimi Heft

ISBN 13: 978-0-13-477760-3
ISBN 10:   0-13-477760-3

1 17

For my family: Pey, Leah, and Bea

Thanks

Writing a second edition turned out to be almost as complex as writing the first. Many people have worked long, hard hours to bring this book to life.

My family—Pey, Leah, Bea, and my parents—put up with me while I was writing it, gave me ideas and inspiration, and made room for this in our lives.

The team at Peachpit has been fantastic. Nancy Davis was enthusiastic about this project from the start and helped me over some of the road-blocks that cropped up along the way. Victor Gavenda helped develop and edit the second edition with patience, a keen eye, and kind humor. Dan Foster was an amazing copy editor, elevating everything that he touched. Danielle Foster and Tracey Croom did a terrific job updating the design while keeping it simple.

My colleagues at cxpartners made a huge contribution to the writing of this book, in particular by patiently bashing the ideas into shape, but I must give special mentions to Richard Caddick and Fiz Yazdi, who covered for me when I needed to focus on writing.

Advice and support from friends, acquaintances, and other authors was invaluable as I stepped outside my comfort zone. Thank you Steve Krug, Kevlin Henney, Yang-May Ooi, Jason Cranford Teague, Louis Rosenfeld, and Caroline Jarrett; and Whitney Quesenbery, and Naomi Pearce; Ken Case of Omni Group; Rich Siegel of Bare Bones Software; Jürgen Schweizer at Cultured Code; Keith Lang, Barney Kirby, Mariana Cavalcanti, Bill Schallenberg, Luis Babicek, Ken Kellogg, Fran Dattilo, and all the folks at Marriott; Alan Colville, David Jarvis, and Pete Greenwood; Tyler Tate, Bonny Colville-Hyde, Dot Pinkney, Jon Tan, Donna Spencer, Dave Ellender, Ian Fenn, Matthew Keeler, Brenda Bazylewski, Alberta Soranza, and the dozens of other people who've contributed, helped, and inspired me.

Contents

Part 3

Strategies for simplicity

Part 4

Remove

Part 5

Organize

Part 6

Hide

Part 1

Why are
we here?

A story about simplicity

The first printer I bought was a fussy device. Setting it up involved fitting together several parts and going on an extra trip into town because the correct cable wasn't included. When I returned, I had to read my computer's manual to check some hardware settings, open up the printer case, and use a paperclip to set some switches. After a few tries I got it right. Then I had to install software onto the computer. The whole process took hours of mistakes, cursing, and painstaking work.

The same could be said of any number of encounters with technology over the years: setting up a mobile phone, plugging a laptop into a board-room display, or reading the weather forecast on a webpage that takes three screenfuls of information and 113 links. Technology that is supposed to make our lives easier often feels like it's rising up against us.

This year I bought a new printer for my home. The setup process was: Take it out of the box, remove the orange sticky tape that was holding the delicate parts in place, pop in the cartridge, and switch it on. At this point the printer informed me that it would like to join my WiFi network and could it have a password, please? And that was it. The printer and my computer got along just fine. Setting up a new printer seemed as simple as plugging in a toaster.

It left me thinking: Why can't it always be like this?

It wasn't the first time I'd asked that question. I've spent my career trying to make technology simple. The problem is that a lot of advice on simplicity is rather vague: "less is more" and all that. So, I've tried to find some strategies that seem to work, and I have real examples and stories to share.

Why should setting up
a printer be any harder
than plugging it in?

The power of simplicity

In 2007, Jonathan Kaplan and Ariel Braunstein turned the US camcorder market on its head by creating a camcorder that was simpler than anything else on the market.

At the time, companies like Sony and Panasonic were trying to win sales by adding advanced features such as the ability to add Hollywood-style captions and video effects in the camera.

By comparison, the Flip was crude, with low resolution and missing "basic" features like optical zoom. One year later the Flip had come from nowhere to sell a million units—at a time when the entire US market was just 6 million units.

Kaplan and Braunstein realized that camcorders had become complex, bulky, and intimidating. Most people didn't want to produce feature films at home—they wanted to pull out a camera, capture a spontaneous event, and share it on YouTube.

The Flip made that as simple as possible, ditching any features that were not essential. There were no cables that could get lost, just a flip-out USB connector that gave the camcorder its name. There were only nine buttons, including a big red Record button. There wasn't even a CD of software for your computer; the necessary software was stored on the camcorder itself, and you could download it when you first connected the Flip to your computer.

Simple products, like the Flip, the original VW Beetle, and Twitter, often have a profound effect on markets. They are easy to use, so they find a popular audience; they are reliable, so people develop an attachment to them; and they are adaptable, so they end up being used in surprising ways.

Today, the Flip has gone, replaced by the camcorder app on people's smartphones. There's no need for a separate device to view videos or upload them to YouTube. Simplicity won again. Simplicity is disruptive; anyone can be caught out by it, even the simplifiers.

**Simplicity is a highly
disruptive strategy.**

Increasing complexity is unsustainable

We crave simplicity, but we're fascinated by complex products. There's something seductive about intricate mechanisms, banks of switches, and Rube Goldberg machines.

Back in 2006, technology columnist David Pogue dubbed this the "Sport Utility Principle: People like to surround themselves with unnecessary power."

His analogy has some depth. At the time, the US auto industry was based on building and selling cars that were big, heavy, expensive, thirsty, and sold at a premium. The auto manufacturers quickly became reliant on selling extras. Then came the economic crash of 2008. Suddenly, no one wanted that unnecessary power. The auto companies found they had driven down a blind alley and that it was going to take years and billions of dollars to make things right.

Continually adding features to software turns out to be equally unsustainable.

Customers will ask for more features. Stakeholders also see it as a way to freshen the product. Sometimes they're right. But the more features you let in, the less chance you have of focusing on what is of real value to your customers. Sooner or later, your new features are going to fall flat. Adding complexity also means you're building a massive legacy of code that makes your product more expensive to maintain, which also makes it hard to react to changes in the market.

Meanwhile, your users become subtly dissatisfied with your product. The added complexity means they can't easily find the features that are important to them. They start to resent the fact that they're paying for features they don't use. They feel intimidated by features they don't understand.

Being fascinated by complexity doesn't mean people like it when they get it. Like the car giants in 2008, you may find that, despite their appetite for more, your users quickly turn against you.

All that unnecessary
power comes
at a price.

Fake simplicity

When something is simple, it looks effortless. So it is always disheartening to discover how hard it is to achieve simplicity. Surely there must be an easier way to reach the goal?

You'll find people pushing ideas to deliver fake simplicity in the form of instructions and guides to help people get to know the system. Like diet pills, laser sights for golf clubs, and "get rich while working from home" schemes, fake simplicity never lives up to the initial promise. Instead, it ends up making things more complex and less effective.

But, remarkably, fake simplicity has become received wisdom. It's a collection of techniques that are quick, relatively cheap, and uncontroversial.

Because of that, you'll find that whenever things get hard, these ideas crop up.

And because everyone "knows" these things work, no one ever gets blamed when they fail.

Instead, people use fake simplicity to say "I'm trying" to the world without ever having to try very hard or be very good.

Instructions and labels seem to say, "See how much effort we've made to explain this to you? If you don't get it, it's your own fault." So they're a great way of faking it, because they shift responsibility for failure onto the user. The problem is that most people don't bother reading instructions; they prefer to get on with doing.

Animated characters who can predict the users' needs and tell them what to do are a more extreme example of instructions. The theory is that hearing instructions from a character will feel friendly and human. But the computers can't accurately predict your needs or tell if you're becoming annoyed with them. Seeing a message in a suggestion box onscreen is one thing. Being told what to do by a cartoon character is another.

Wizards promise to make things simple by breaking them down into steps. The problem is they take control away from the user. Because of this, wizards feel constricting. It may be possible to herd users through a brief wizard, but the longer it goes on, the worse it feels.

Sticking on these kinds of extras rarely makes an experience feel simple; they're just another thing for the user to deal with.

Password help

Your Password:
- Must be 8-32 characters long
- Must include at least two of the following elements:
    - At least one letter (upper or lower case)
    - At least one number
    - At least one special character from the following:
      # $ % ' ^ , ( ) * + . : | ? @ /] [_ ` { } \ ! ; - ~
- Must be different to your previous five passwords
- Must not match your User ID
- Must not include more than 2 identical characters
  (for example: 111 or aaa)
- Must not include more than 2 consecutive characters
  (for example 123 or abc)
- Must not use the name of the company
- Must not be a commonly used password
  (for example: password1)

OK

**Instructions
alone don't make
things simple.**

Things fall apart

In the face of complexity, it's easy to ask: Why can't the users just learn? In return for a little attention, they will have access to so much power. And often, "the users" have asked for that power in the first place, so they surely bear some responsibility, too. Personal experience should tell us why that argument it wrong.

Take the Print dialog you find in most software at the moment. It contains most of the options you'd want when printing a document—setting the quality, setting the quantity, choosing whether to print double-sided, and so on. Over time, we've all learned what each of those features does. We put in time and effort in exchange for power and flexibility.

Most of us commit documents to print at the last minute when the pressure is on: The deadline is looming, the courier is waiting, the meeting guests are arriving. Magically, the printer breaks down.

It's not magic, it's us.

Learning breaks down under pressure. The more complex or rarely used the knowledge, the faster we lose it. This is why technology fails just when we need it most: We press the wrong button, miss the vital step, or enter the wrong code without realizing it.

Experts who use complex systems are trained to slow themselves down at critical moments. Pilots are taught to read checklists; Japanese train drivers are taught to point at controls and say out loud what they're about to do.

It turns out that the price of complexity is not only learning; it's also the awareness of the need to slow down under pressure. That behavior requires experience and yet more training—it's not part of everyday behavior. It is inevitable that your users will hurry, will get distracted, and will forget what they've learned.

When your users are under the most pressure, they need simplicity most.

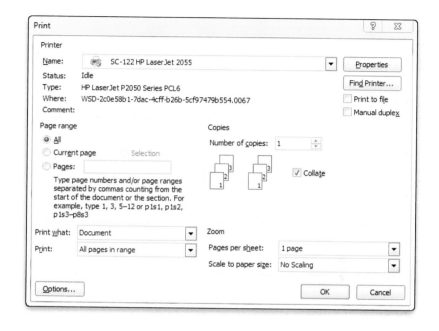

Expecting the user to learn is unsustainable.

Elegant simplicity

One of my favorite quotes about simplicity is attributed to Oliver Wendell Holmes Junior (who served as a US Supreme Court Justice in the early 20th century). He said:

"I would not give a fig for the simplicity on this side of complexity; I would give my right arm for the simplicity on the far side of complexity."

In other words, there are two kinds of simplicity.

The first kind comes early on in our exploration of a problem. Our knowledge is incomplete, our ideas simplistic, we oversimplify, and we lose something important.

But as we get into a problem, we discover how complex it is. There are questions, subtleties, and interdependencies that we hadn't expected. At this point, our solutions are also complex and hard to understand without instructions.

At some point, we start to notice the deeper, underlying patterns and see a different kind of simplicity—one that manages to pack all of the complexity of the universe into a few equations. The complexity remains, ready to be unpacked, but we are able to contain it within this elegant simplicity.

When you set simplicity as a goal, different people hear different things.

Some people understand the problem well; they'll assume that getting to a solution will be quick and cheap. "There's no need to overcomplicate things," they'll say. They may sound like allies, but they'll abandon you as soon as the real complexity starts to emerge.

Some will complain that you've failed to see things as complex as they truly are. They will raise excellent objections to any attempt to find a simple solution.

A few will back you to find the deeper simplicity, but they need to know that the rewards of elegant simplicity don't come easily.

When you set out on the journey to simplicity, it's important to begin by describing those three reactions. If you take time to prepare people for what's ahead, they're more likely to back you to the end.

There are two types of simplicity, according to Oliver Wendell Holmes Jr.

Not that kind of simple

I was once called in to review a company's new business intranet. It had recently been redesigned, but the salespeople complained that it was making their work impossibly complex.

The salespeople showed me how they had to fill in page after page of forms every time they met a potential client. I was puzzled why such a bureaucratic system had been put in place.

Then I talked to the managers who had set up the intranet. They told me how wonderful the new intranet was and how much time and effort it was saving them because it "automatically" generated the reports they needed.

Sure enough, the reports exactly matched the forms the salespeople now had to complete. The managers had made their lives considerably simpler—by making the salespeople's lives more complicated.

When you're designing any piece of technology, there are at least three perspectives: the manager's, the engineer's, and the user's.

This book is about the user's perspective: it's about making things feel simple to use.

Sometimes you can create simple user experiences with simple technology, or simple management, but that's not a certainty. Google deploys complex technology and employs thousands of people to make it easy to find information on the Internet.

What feels simple to one person in one situation may not feel simple to everyone in every situation. A Formula One driver won't feel his life has been made simpler if you ask him to race in a Mini. But while it's a fun puzzle to design complex systems for experienced users, technology becomes interesting when it gets out of the hands of experts and finds a wider audience.

This book is concerned primarily with the experience of mainstream users.

Simpler than a bike.
Until you try to ride it.

Character

Simple doesn't mean minimal. Stripped-down designs can still have their own character and personality.

Take two simple chairs: a Shaker chair and an Eames chair. Each reduces the chair to its basic components. Each is easy to manufacture, given the technology available at the time it was designed. And each solves a different problem: The Shaker chair is hard-wearing; the Eames chair is light and comfortable.

The two designs are simple, yet they have utterly distinctive characters that derive from subtle differences in their purposes and technologies.

The materials you use, the emphasis you place on the key elements, and the way you combine even a few elements will have a dramatic effect on the final design. People will recognize and put value on the small differences, just as they focus on the small differences between Google and Bing searches or between one online bank and another.

> Simplicity does not mean want or poverty. It does not mean the absence of any decor, or absolute nudity. It only means that the decor should belong intimately to the design proper, and that anything foreign to it should be taken away.
>
> —*Paul Jacques Grillo* (Form, Function & Design)

In other words, you can be simple without being minimalist. The character and personality should come from the medium you're using, the brand you're representing, and the task users are undertaking.

Both simple. But each has a unique character.

Be single-minded

Simplicity has its origins in single-mindedness. When you're clear about what you're trying to achieve, it becomes easier to see when you're going astray.

But organizations contain many points of view, all of them held by reasonable people who think deeply about their work. It can seem as though organizations have an immune response to making things simple.

A few years ago, I spoke to a manager at an automotive company who'd been tasked with simplifying their product range. Every time he tried to cut an option, he'd get a complaint from one of the salespeople: "That option is vital to one of my customers. Even if the customer provided a tiny percentage of the company's entire business, the salesperson would point out, "Well, they're my most important client."

The problem was that the organization lacked a single-minded strategy. Optimizing for individual customers ("all customers are important") is a valid strategy—so long as customers are asked to pay for the additional effort required. Optimizing for a simple product line is also a valid strategy—so long as you're willing to ignore some customers' specialist needs.

If you're pursuing simplicity, you need an appreciation of how to create the feeling of simplicity in the users' mind. But before you can do that, you need a single-minded vision.

Achieving this requires us to resolve many different points of view. Which customers should we serve? What problems should we solve? Should we build a solid technical foundation for tomorrow or respond to the needs users have today? Resolving the conflicting voices of stakeholders, and the conflicting voices in your head, requires careful attention.

Simplicity begins with single-mindedness.

Part 2

Setting a vision

Making sense of the muddle

Whether you're designing a multichannel service or a drop-down menu, you need a way of judging whether you're keeping things simple. A vision brings clarity when complexity starts to crowd in.

That sense of complexity arises when you're unable to prioritize many competing ideas. The purpose of a good vision is to make it obvious what you must do.

When I find myself tied in knots over a design, I take a step back and ask myself, "What is the user *really* trying to do here?" The answer becomes my north star for a simple design. This works especially well when I am designing something very small (like one page in a larger website) and when I know more or less what I have to design.

The way I ask that question is important. I could ask, "What am I trying to do here?" In other words, "What is it that I'm designing?" That question brings some clarity and will simplify my design. It's a good start.

However, the focus is on what I'm doing. It's the user who matters, and it's the user's sense of simplicity I care about. So it's better to ask, "What is the *user* trying to do here?" For instance, if the user is trying to register for a website, then my job is to make that feel simple.

But going a little deeper and asking, "What is the user *really* trying to do here?" is a powerful question because it makes me ask "*why*?"

Is the user trying to register because they want to protect something they value (such as their bank details)? Is it because they want to get a reward you're offering them (which is a weaker reason)? Is it because we're *making* them do it, and it's a barrier we've put in their way?

No matter how big or small the problem, your vision should focus on the user and help you understand what is happening and why.

The starting point is often a tangle of ideas.

Alignment

If you're designing for an organization, then other people, stakeholders, will be involved. Often, they will pull in different directions creating friction that slows progress. If things get bad enough, the project will fail.

To keep things moving, it's tempting to allow each one to have a say on the design. This shows that you're listening and keeps them onboard. In truth, no one likes this game. Everyone despises "design by committee." Everyone knows the result is always a muddle of ideas and compromises that is hard to fathom—the opposite of simplicity.

Nevertheless, listening and accommodating stakeholders' constraints makes for a better, sustainable design. Stakeholders have important agendas and real problems to tackle.

They must live with the consequences of your design. We all dream about designing as a dictator, but this is childish fantasy.

Don't seek agreement through bartering and compromise; seek it through alignment. If everyone feels they are working toward a common goal, friction eases, and the right solution becomes clear to all.

While stakeholders have personal and professional agendas, there is one point of view on which they should all align: doing what's right for the organization's end user.

Marketing wants to promote the product, to create happy users. Legal wants to ensure compliance, to avoid litigious users. Technology wants to develop robust solutions, to create satisfied users. The design is where their agendas meet the user.

The answer, then, is to align them around a vision based on the end user's real needs. Just don't expect them to recognize those needs easily.

Stakeholders spend so much time living in their silos that their views of the end user can become twisted to suit their agendas. They're not willfully wrong; they are blinded by circumstances. And you will be blinded, too. Everyone needs to open their eyes.

So bring stakeholders to meet the end user and discover the vision for themselves. Do this, and the friction and politics will ease.

**No one likes design
by committee.**

Get out of your office

Most designs are reviewed in quiet meeting rooms where everyone gives the design their full attention. People rarely use designs in such a calm setting. Even at work, they are constantly interrupted, uncovering new information, and changing their minds. Simple user experiences need to work in chaotic environments.

If you want to set a vision, begin by visiting the place where people will use your design. Often it will transform your understanding of the problem.

A few years ago, I was asked to redesign some software to help car dealers put together marketing plans. The brief was to merge several components into one so that a dealer could write a plan in one sitting.

Fortunately, a colleague of mine visited some dealerships to talk to the managers about their needs. At the first dealership she visited, the manager sat in an office with a glass front that opened onto the showroom. As they spoke, the manager kept glancing up to scan the showroom. Whenever a customer looked lost, he would hurry out and attend to them. It was the same in every dealership she visited: The managers were constantly interrupted by the needs of their customers.

Instead of merging the components, we needed to break them into smaller chunks so that the managers could complete them in the short bursts of time they had.

If we'd simply imagined the managers at their desks, we would have missed this crucial detail.

Even with minimal planning, you can learn a lot from a few hours spent watching users in their everyday environment.

If you can't get permission to do it, then talk to some users about where they are and what's happening when they use your software.

I once reviewed a mobile website that had been promoted during a rugby tournament. The owners couldn't understand why users dropped out of the site after a couple minutes. Their exit points didn't correspond to any obvious usability problems.

When I interviewed users, the answer became clear: They had been using the site during the ad breaks. When the rugby came back on, they went back to watching TV. The site took too long to get through.

You can't control the environments where people use your software. You must design it to fit.

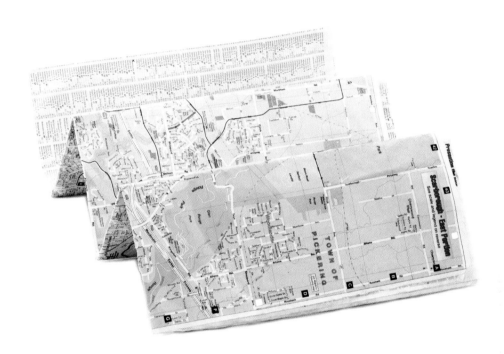

The best place to watch users is in their natural environment.

What to look for

When you get into the real world, you'll notice lots of ways that people's experience can be affected. Here are some things to be ready for:

Offices

- In open-plan offices, staff frequently distract each other. Watch, and you'll be surprised how often people are interrupted or drop what they're doing because they've overheard something interesting.

- Instant messaging, alerts, and email interrupt users constantly.

- The physical arrangement of the space affects how easy it is for certain people to interact and the styles of interaction (meeting, short discussion, water cooler chat).

Homes

- People use their devices while watching television or listening to the radio, with their attention and time divided unpredictably between the two.

- People often play different social roles at home: the organizer, the ideas person, the authority, the disruptor. These can depend more on task than stereotypes.

- Where does the family organize and share information? In piles of correspondence? On the refrigerator door? In filing cabinets?

Outdoors

- Stand on a busy street, and you'll see people steering around each other as they use their devices to get directions or send messages.

- People may be carrying bags while they try to use their mobile phones, making it harder for them to tap on small buttons.

- People check mobile apps in lines everywhere—they may be interrupted at any time.

- Bright sunlight can make it hard to read mobile screens outdoors.

- Larger devices, such as tablets, quickly start to feel heavy and uncomfortable, making people want to put them down.

Your user experience needs to be simple enough to work among the distractions and fit into the cracks between interruptions.

DO NOT
DISTURB

NO MOLESTE

PRIÈRE DE NE
PAS DÉRANGER

BITTE NICHT
STÖREN

At home, at work, and outdoors, you must design for constant interruptions.

Three types of users

When it comes to simplicity, you can divide users into three types.

Experts are happy to explore your product or service and to push the limits of what it can do. They want never-before-seen technology customized for them. Even if they're new to a product, they have an expert attitude. In other words, they'll spend time finding out how it works and exploring new features. If you're making a mobile phone, these are the people who want to be able to browse through the mobile phone's file system and tweak everything. It turns out there are relatively few people like this.

I call the next group **willing adopters**. They probably already use some similar products or services. They're tempted to use something more sophisticated, but they're not comfortable playing with something entirely new and must be given easy ways to adopt new features. For instance, they might be interested in a more sophisticated phone, but only if they can transfer their precious contacts easily. There are fewer of these people than you'd imagine, and their tolerance for learning is low.

The vast majority of people are **mainstreamers**. They don't use technology for its own sake; they use it to get a job done. They tend to learn a few key features and never add to their repertoire. These are the people who say, "I just want my mobile phone to work." Most people fall into this group.

It's tempting to think that after a while, people graduate from one group to another. But that hardly ever happens. Even after years of using a product, people tend to stay in the same group.

For example, take any large group of people who've been using Microsoft Excel for five years. You'll find some people who've explored settings and options, some who've got a few specialist features set up to do what they like, and others who just use it for adding up columns of figures.

It has more to do with their underlying attitude toward technology than the amount of time they spend using a product or service.

It's tempting to design for the first two groups—they're easier to please. But experiences that feel simple are designed for the mainstreamers.

The vast majority of users are mainstreamers; experts and willing adopters are a minority. For example, in 2009, complex cameras like SLRs made up only 9 percent of the digital camera market (source: CIPA).

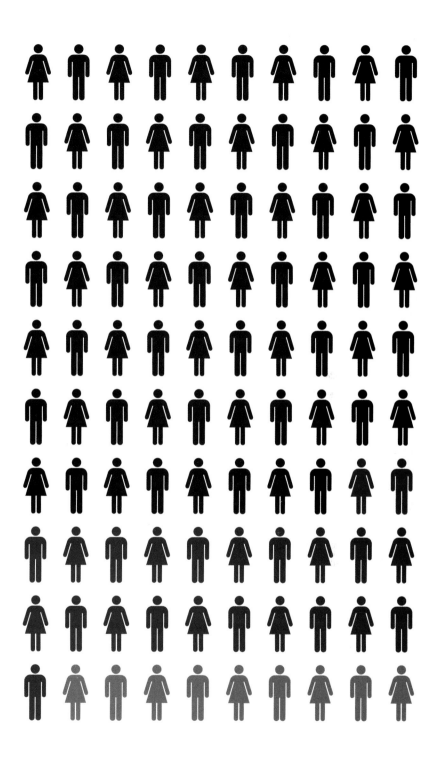

Why you should ignore expert customers

Most companies spend too much time listening to their expert customers—those who spend the most time using their products or services—because they're easy to talk to. Expert customers are enthusiasts; they're vocal and opinionated about how to improve what's offered.

But experts aren't typical customers, and their judgment is often skewed. They don't experience the problems that mainstream customers have.

And they want things that mainstream customers don't care about.

Here's what one responder on Slashdot (a blog run by experts and enthusiasts) had to say when the iPod was announced: "No wireless. Less space than a Nomad. Lame."

Another commenter wrote: "I don't see many sales in the future of iPod." And another: "All I can say is, as an Apple 'fan,' I'm sad."

Commenters on another enthusiast blog, MacRumors, also wanted more: "I still can't believe this! All this hype for something so ridiculous! Who cares about an MP3 player?"

Apple's expert customers wanted a flying car. Apple's mainstream customers just wanted an MP3 player that worked.

I see this again and again: A small group of customers makes noisy, persistent demands for new features that are too complicated for typical users.

You'll find it hard to convince your stakeholders (who are insiders and therefore experts) that the customers who are also experts (just like them) are not the ones you should listen to. After all, your best customers spend a lot of time and money per head; they're so easy to talk to—they come to you, they get what you do, and they speak your language; and they're so reasonable—if you ask them to upgrade to the latest version, they'll do so without hesitating.

But if you listen to them first, you'll create products that are too complex for mainstream customers to use.

By January 2010, Apple had sold 240,000,000 iPods and no flying cars.

So if your stakeholders are trying to create a mass-market product by listening to their expert customers, remind them of this story. Sometimes, it's best to ignore your expert customers.

**Experts often want
features that would
horrify mainstreamers.**

Design for the mainstream

The middle ground looks safer. Unlike the demanding enthusiasts, the willing adopters would like to use some fancy new features, as long as you make them just a bit easier.

Most "usable" design tends to focus on this group. People who already book their flights online are invited to user tests for travel websites. People who already use the camera on their mobile phone are asked to test camera phones. So we design for people who aren't very hard to please.

You can learn a lot by watching these people. Every user test I've watched revealed some way to improve a website or a mobile phone. But by focusing on these people, we're making it easy on ourselves.

These users will put up with the problems they've grown used to (like needing to dig around on their mobile phone to find their photos) because they've learned to tolerate them.

But these willing adopters are still not typical. They're a small, extreme group who have more skills and more perseverance than mainstream users. It's just that they're a bit less extreme than the experts.

If you want simplicity, if you want to be seen as an innovator, then it's the mainstream customers you should be aiming at. The Ford Model T wasn't the first car ever built, but it was the first one made with the mass market in mind. Henry Ford revolutionized the motor industry because he aimed squarely at the typical person. Simplicity was at the heart of his vision:

> We will build a motor car for the great multitude. It will be...small enough for the individual to run and care for. It will be constructed...after the simplest designs modern engineering can devise. But it will be so low in price that no man making a good salary will be unable to own one.
>
> —*Henry Ford, on the Model T*

All of Ford's innovations (his use of production lines, the price of his car, the easy-to-maintain engine design) came as a result of his desire to focus on creating a simple product that was suitable for the mainstream.

If you want to make something simple, design for the multitude.

If designing for experts is like building a car for mechanics, then designing for the middle ground is like building one for people who like tinkering with engines. The typical user is a mainstreamer.

Mass appeal comes from focusing on the mainstream.

What mainstreamers want

When you're setting your vision, make sure the mainstreamer is at the heart of it so you can't sneak in the convenient skills of the expert to get you out of a tricky design problem.

- Mainstreamers are interested in getting the job done now; experts are interested in customizing their settings first.

- Mainstreamers value ease of control; experts value precision of control.

- Mainstreamers want reliable results; experts want perfect results.

- Mainstreamers are afraid of breaking something; experts want to take things apart to see how they work.

- Mainstreamers want a good match; experts want an exact match.

- Mainstreamers want examples and stories; experts want principles.

Don't assume you can teach users much or that instructions will help them. When they're under pressure, mainstreamers tend to forget what they've learned, ignore instructions, and revert to behaving like novices.

You've probably experienced that for yourself: When you've got a deadline or when you're distracted, that's when you delete your vital file or the printer starts spewing out the wrong document.

Simple user experiences need to work for a novice, or a mainstreamer who's under pressure.

Mainstreamers don't want to build products from scratch.

Deeper needs

Asking "What is the user *really* trying to do here?" means you uncover users' deeper needs.

Jürgen Schweizer is one of the developers of Things, an award-winning iPhone "to-do list" app. He points out that understanding what your design should do is often trivial: "At first glance, a to-do list is just a list of items with a checkbox next to each one so the user can see what's been completed."

However, even with something as straightforward as a to-do list, they want to use the app for a reason—they have a deeper, emotional need, says Jürgen:

> When we thought about why people would use our software, we realized that they had a lot on their plate. They wanted to achieve a lot and still feel in control. They needed to be able to capture a thousand items and yet not feel overwhelmed when they looked at the list. So we put a lot of effort into making sure that they'd only ever look at a handful of the most important things, but they'd be able to find all their other notes and reminders just when they needed them.

To-do lists can easily get out of control. What starts off feeling simple can quickly feel complex. But Schweizer didn't want the app to annoy users by interrupting them with demands that they organize their lists.

That insight led the Things team to spend time finding natural and helpful ways to organize and filter tasks. As Schweizer points out, getting that right was a very subtle and complex problem. "It turned out to be about making the user feel good about putting things off. We needed to make the user feel confident that they'd be able to put tasks away and find them again later." Solving that problem is what made Things stand out from hundreds of other iPhone task managers to become a popular and enduring app.

The time spent discussing those deeper, emotional needs helped the developers of Things understand the real reason people needed their software and made them focus on an important set of hidden needs.

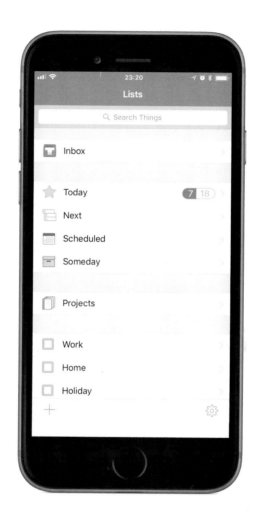

**Even a to-do
list must satisfy
emotional needs.**

Branding simplicity

If you're designing an experience for an organization, the feeling you leave someone with should fit the organization's brand. If the organization prides itself on being "energetic," then your design should have a dynamic quality. If the organization stands for "reliability," then your design should make them feel secure. When things fit our expectations in this way, they feel simple.

Organizations often have rather confusing descriptions of their brands, and many have no description at all. Yet people within organizations often sense when something is off-brand, even if they can't explain why.

In those situations, it helps to have a simple framework to try to think about what a brand should be. Here's one I like to use.

You can think of a brand as having three qualities:

- Practical—what does this brand do to help me?

- Emotional—how does this brand make me feel?

- Ethical—what does this brand stand for? (Ethical qualities don't have to be high-minded: Some beliefs are quite frivolous.)

When you know the answers to these questions, it helps you to understand what parts of the experience to emphasize in your design.

For instance, take a brand like Virgin Atlantic. It's clear that the practical quality is: making flying simpler. That may seem obvious, but you'd be surprised if they started selling power tools. For most brands, the practical benefit is often to make something simpler, which is useful validation for a strategy based on simplicity.

The emotional quality is: fun. So you'd expect the brand to have a playful quality (in interactions where playfulness was appropriate).

And Virgin Atlantic's "ethical" stance is to be the rebel that thumbs its nose at the conservative airlines. So you'd expect the brand to make fun of competitors.

The trick is to include just enough character, without forgetting the basic simplicity you're aiming for (think of those characterful yet simple chairs from the previous chapter).

Those three bullets give you clues that help you set a vision of designing something that is both simple and characterful.

The brand is not the logo—it's the experience.

Simplicity is about control

Unraveling your users' emotional needs can be tricky and made worse by the fact that many people are uncomfortable sitting in a design meeting talking about feelings.

Fortunately, when it comes to designing for simplicity, the key emotional need is for users to feel that they're in control.

First, they want to feel in control of the technology they're using.

Experts want to control and customize the technology. You'll need to take the mainstreamers' view of "control:" to be in control of the outcomes. They don't want to worry about the software or technology, and they don't want it to tell them what to do. Mainstreamers want control that is easy, reliable, and quick.

Your design shouldn't interfere with this sense of control. It should extend it. Simple experiences make users confident that they're making good choices. Simple experiences reassure users that the product will respond in a predictable way.

Second, they want to feel in control of their lives.

Sometimes being in control is about completing a task: A woman buying a dress wants to feel in control of how she looks. Sometimes it's about getting information: A man reading the news wants to understand what's happening in his world (in order to feel in control).

Begin with that need—the user's need to feel in control of some part of his life—and then try to dig deeper by asking, "So what?"

Take the example of the Things app from the previous section. The users' overall need is to be in control. So what? If it's useful, they will write down lots of tasks. So what? Having too many tasks on their to-do list means they'll feel overwhelmed. So what? They need to be able to limit what they see to what's relevant at any one time. So what? We need to come up with an easy way to organize their lists.

Repeatedly asking "So what?" eventually throws up an emotional need, a rational need, and a solution. It also helps you arrive at a deeper understanding of the design problem you want to solve. (Of course, you'll need to check your thinking by talking to real users.)

Once you understand who your users are and what drives them, you'll have some of your most important insights.

**Simplicity is about
feeling in control.**

Choosing the right "what"

Often, designs feel complex because they allow unimportant steps to crowd out what is at the core of a design, or they ignore the steps that provide context to the design. When you're setting a vision, it's these steps that help you simplify the experience.

Look at a "professional" camera, and you'll see myriad controls that can make it hard to spot the core button: the one that takes the picture. Making this action obvious and easy is important if you want a camera to feel simple to mainstream users.

But the context of taking a picture is what comes before and after you press that button. When you're about to take a picture, you see something in the moment and you need to be able to find and activate your camera immediately. When you're done, you want to be able to share the picture easily.

These "context" steps are as important to creating a simple experience as the "core" step.

The thing to do, then, is to describe the core and the context in a way that tells the story from beginning to end. What you're leaving out is all the fiddling, organizing, optimizing, and switching around.

Remember that it's the user's actions you're most interested in, not the thing you're designing. If you describe your solution in too much detail at this stage, you may end up painting yourself into a corner. Instead, just go to a level of detail that's sufficient to tell the story. You might start with "Quickly take and share photographs" to begin with and then list each step the user will take, keeping a consistent level of detail.

Make sure you describe what's happening in the user's language or you risk losing track of what's core. People who use Facebook aren't "social networking"; they're sharing pictures and news with friends. If you get away from describing things as the user sees them, you'll end up writing a story about a database or a mobile phone instead of about the user.

Focus on the main action and describe it as the user sees it.

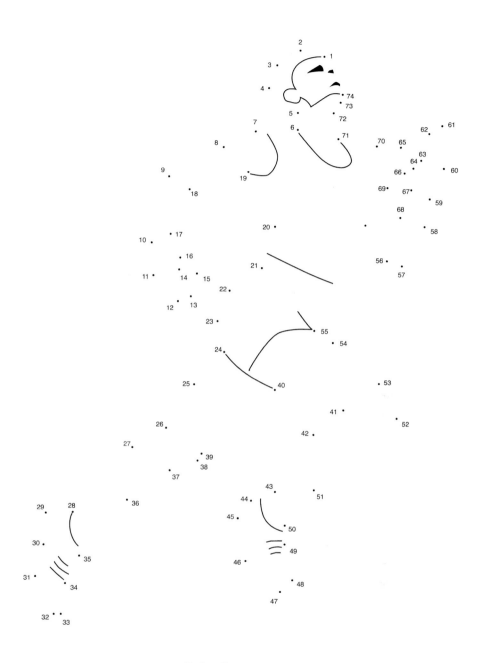

Make sure you don't miss any important steps.

Describing the user experience

Once you've researched your problem, you need to turn it into a vision. A story is a great way to describe your vision. Unlike a list of requirements, it helps the reader understand what's important and why.

In case you're worried that stories aren't very businesslike or technical, don't be. Managers use stories all the time (mission statements, for instance), and technical teams tell stories (flowcharts and use cases). User experience teams have been writing stories for a long time, too.

A story should sum up the core experience in a few sentences. For a video app, it could be:

> You're on a city street when you hear a commotion: Jennifer Lawrence is walking toward you. You pull your phone out of your pocket and hand it to a stranger, asking her to video you with Jennifer in the background. Then you share the clip on Facebook.

If you're trying to design a video app, this story tells you what's important:

- It starts up quickly (because Jennifer isn't stopping for you), and it's simple enough that someone who has never seen it before can use it immediately.

- It's easy to find and edit clips.

- And, finally, the purpose of taking videos is to share them.

Stories manage to pack a lot of information into a few words. They're efficient. They're also easy to remember and to share, which means they're more likely to come up when you're discussing design decisions. In fact, people love stories so much that if you don't give them a story, they'll invent their own ("If I were using this camera, I would…"), which can drag your vision all over the place—so make sure it's your story they're using.

It's worth spending time to get the details of your story right; if you're designing for simplicity, those details are especially important.

Imagine you're standing on the street when...

Putting it all together

Don't worry too much about the form of your story. What matters is getting your constraints into a format you can share. Sometimes that's just a few words. Other times you might make a storyboard, prototype, or video.

Keep your story minimal. Don't get sucked into describing events in detail. Instead, describe each goal and identify the feature that resolves it (the core features). There are three reasons for this. First, a brief story is easier to remember and retell, so it's more likely to be used. Second, it's easier for people to imagine how that story could play out in other circumstances (so you can imagine handing your phone/camera to a parent at a kid's birthday party). Finally, adding detail into a story is like a movie camera zooming in: People assume that their attention is being drawn to something important. At best, this will feel odd; at worst, people will invent reasons why the detail matters. So only add details that matter and that help you explain the story.

Show, don't tell. We're used to trusting people's actions more than their words. Descriptions of users' behavior will make a stronger impression than assertions about their character. Don't say the protagonist is detail-oriented; mention that she cross-checks her work with her notes. Showing makes something concrete.

Don't invent. Your story needs to be credible, and to be credible it must be based on real people and real events. The Jennifer Lawrence story I told about the camera is based on something that really happened to a friend of mine. Your specific story might combine elements from several events, which is truthful even if it's not a true story. But unless your story is based on real events, you won't be able to back it up, and it will feel artificial. Using relevant details, as described here, makes your story concrete and believable.

Practice it, tell it to someone else, refine it. Doing so will help you find and fix the flaws in your story and help you boil it down to the essentials.

A good user story is brief, concrete, credible, and uses relevant detail.

> "Writing is hard work. A clear sentence is no accident. Very few sentences come out right the first time, or even the third time. Remember this in moments of despair. If you find that writing is hard, it's because it *is* hard."
>
> —*William Zinsser, On Writing Well*

Now tell a story.

World, character, plot

If you look back at the vision we've developed, you'll see there are three levels:

- A believable world (the "where" and "when" of our story)
- Credible characters (the "who" and "why")
- A coherent plot (the "what" and "how")

Many designs feel complex because they don't take into account the pressures of the real world, because they expect the user to be able to cope with anything, or because they miss vital steps. Your design needs to fit comfortably into the complete story.

Michael Johnson, Moving Pictures Group Lead at Pixar, has described how Pixar uses this approach to creating movies. The movie is built from the outside in, starting with a world (toys are alive when people aren't around), adding characters and motives (a cowboy who's jealous of the new spaceman toy), and finally describing the plot (they fight and fall into the hands of a toy killer and must cooperate to escape).

If they run into problems with the plot, they go back to the characters to understand what they would do. If they run into problems with the characters, they look at the world to see how it shapes them.

The same goes for our user experience story for the Flip camcorder. If you want to know how the person taking the movie would act, you need to look at who they are (someone who has never used the camera before) and the world they're in (a crowded street with no time to ask questions), so they'd be flustered and they'd look for one simple button to press.

Place your design within a plot, driven by credible characters and set in a believable world. In the words of architect Eliel Saarinen: "Always design a thing by considering it in its next larger context—a chair in a room, a room in a house, a house in an environment, an environment in a city plan."

World

Character

Plot

Extreme usability

When you look at stories of simple experiences, it's clear what sets *simple* experiences apart: They work under extreme conditions.

To be simple, you must aim for something tougher than the regular goals for usability.

Usability aims for...	Simplicity aims for...
a specific group of people can use it	anyone can use it
easy to use	effortless to use
responds quickly	responds instantly
understood quickly	understood at a glance
works reliably	works always
straightforward error messages	error-free
complete information	just enough information
works in a user test	works in a chaotic environment

Targets like "instant" and "effortless" are intimidating because, in truth, they're unattainable. But there's an important benefit of shooting for a target you can't hit: It keeps you facing in the right direction.

Imagine setting a target of "responds quickly" instead of "responds instantly." It would be easy to justify making a change that would slow down the response time by only a second—after all, that's still "quick," isn't it?

Slowly, with each successive change, you find your design stops being simple and starts becoming slower and more irritating. Compromises like these happen all the time in design meetings, and this is why the products we love often turn into monsters we loathe.

If, instead, you set a target of "instant," you find yourself looking for changes that make the experience quicker.

It's been pointed out that products that start out simple often end up getting so complex they cease to be useful. But if you set extreme targets, over time your product gets better (or at least achieves the goals that really matter).

Aiming for extreme targets, even ones you can't quite reach, will help you keep your product simple.

Designing simple experiences means reaching for extreme targets.

The quick and dirty way

The quick way to get to a vision often works when I'm making minor improvements or working on something small, like a single web page.

I begin by describing what I'm designing in plain language. I say it out loud to anyone who'll listen—I find it comes out better that way. If it sounds odd or the listener doesn't understand what I've just told them, then I know I need to rephrase it and try again, usually with someone new to whom I've not given any clues. If there's no one available, I'll write it down, but explaining it to a person is always best because their reaction tells you whether you're getting it right.

My aim is to come up with a description that is concise, clear, and complete.

I try to make it no more than one short sentence. If I can sum up the main activities without branching off into details or losing the listener's interest, then I know I've made it concise.

If the listener understands it right away, I've probably made it clear.

I don't try to list every feature. I just try to explain the main features at the same level of detail. If I can summarize the main points without leaving out important details, then I know I've made it complete.

For the Flip camcorder, the description is "take and share video." For a newspaper home page, it's "a summary of the most important events right now." Even a complicated device like the iPhone can be described by its core components: Steve Jobs introduced it as "a widescreen iPod...a revolutionary mobile phone, and a breakthrough Internet communications device."

When I've done that, I make sure I set some constraints on the actions to focus me on making them as simple as possible. So for the Flip, that would be "take video instantly and share it effortlessly."

It normally takes a few iterations to get it right, but it always pays off because it helps me focus on what's important.

Describe what you want in the simplest possible terms.

Insight

The magic happens when you take the things that you've learned in researching your story and turn them into a deep understanding of the problem you're trying to solve.

The trick turns out to be simple. It only looks like magic when you've had enough time and practice to make it appear effortless.

- First, look back over the facts you've gathered about your users, the problems they face, and the world they live in. Prioritize the things that have the most impact on your users' behavior. For instance, in my earlier example about the car dealers, interruptions from customers had a huge impact on the dealers' task of creating a marketing plan.

- Next, look for points in your story that you can act on. In the car dealers' example, we couldn't stop the interruptions, but we could make the tasks shorter and help users to pick up where they left off by giving them a checklist to complete.

- Now prioritize those points: Where can you have the most impact? What can you change easily? For the car dealers, creating shorter tasks had the most impact on the dealers' ability to complete their marketing plans, so it became our top priority.

- Finally, test your insights. What would happen if your ideas weren't true? Is anything likely to change that will affect your insight? Can you see examples or counterexamples anywhere already? Do these reveal flaws in your insight or are there other reasons (for instance, a poorly executed design)?

Testing your insights means spending more time watching people in the real world, often using prototypes or competitors' products. This is where you find the small differences that make your insights valuable.

Spend time reviewing the data behind your story and discussing it.

Take time to reflect on your story.

Getting the right vision

Whether you take the long route or the quick and dirty route, writing a vision will take longer than you expect.

"As designers, we want to start designing immediately. It's important to resist that," says Jürgen Schweizer of Cultured Code. Starting design early means missing out on important insights. It can even mean designing the wrong thing entirely.

A few years ago, an automotive manufacturer asked me to design a car selector. They already had a design in mind: Make it easy for people to choose a car by asking them questions about their lifestyle and personality and then offer a short list.

When I tried out the idea on customers, they told me that their answers would be lies. "If I tell them that I have a dog, they won't let me see a sports model," one customer explained. Customers quickly became irritated by the convoluted process of finding a car by describing their hobbies.

It turned out that customers had a general idea of what they wanted. They could make a choice they were happy with if they were just given clear photos of a lineup of cars.

Spending time understanding the problem leads to better, simpler solutions.

> "When you start looking at a problem and it seems really simple, you don't really understand the complexity of the problem. Then you get into the problem, and you see that it's really complicated, and you come up with all these convoluted solutions. That's sort of the middle, and that's where most people stop.... But the really great person will keep on going and find the key, the underlying principle of the problem—and come up with an elegant, really beautiful solution that works."
>
> —Steve Jobs (*quoted in* Insanely Great: The Life and Times of Macintosh, the Computer that Changed Everything *by Steven Levy*)

As Luke Wroblewski, Product Director at Google, says, "Your first design may seem like a solution, but it is usually just an early definition of the problem you are trying to solve."

In my experience, roughly the first third of any project is spent trying to figure out what's really important. It's a nerve-wracking time, as complexity seems to spiral and there's no solution in sight. Sticking with it is the first and most important step in achieving simplicity.

Don't rush into design. Understanding what's core takes time.

The really great person
will keep on going...
and come up with an
elegant, really beautiful
solution that works.

—Steve Jobs

Share it

In 2002, Alan Colville was a product manager at Telewest, a British cable TV company. He'd been charged with upgrading the set-top box software, a job that touched on every part of the company's workforce, from software developers to call centers. As he described it:

> People at the company were pretty cynical about new projects, and change was seen as a bad thing. Everything we'd done before was too complex, had needed fixing after it was released, and was irritatingly slow. We needed to show people that this project was going to be different in that it focused on our typical customers and their needs. Bringing this new focus, we wanted to deliver something that was the opposite of what we'd done in the past by being simple, stable, and fast.

Colville started putting up posters around the company, promising that the project was going to make the set-top box "simple, stable, fast."

Those three words became the guiding principles for every decision: "Will it make the experience simpler, stabler, faster?" was a question that he asked at every meeting. Colville remembers:

> I knew it was working when I was on a conference call and a project manager was telling me about an idea that had been dropped. She told me, "It would have made it simpler and stabler, but not faster—so we're not going ahead with it."

> The stress just fell away and the design started to go right. Normally the company would hemorrhage money to customer support whenever there was a new software release. This time, when we released the software, our support call volume was negligible. We saved £3 million on that alone.

Sometimes your vision is as neat as three words. But Colville says his favorite tool for sharing a vision is a quick prototype of the finished product.

Sharing your vision means that the right decisions get made even when you're not there. It means your stakeholders can tell the difference between good decisions and bad decisions.

Making your core statement visible reminds people how important it is. Using it all the time makes it second nature to them. Putting it in the public eye means everyone on the team knows they must deliver what's expected of them.

Repeat your story to everyone involved with the project, every time you meet them. Don't stop retelling your story. When you're getting bored of it, the message is just starting to get through.

Tell your story.

Part 3

Strategies for simplicity

The change curve

Simplifying something means change, and change always involves pain. Most people put off pain at all costs. We all like to hope that things will get better tomorrow, rather than making painful changes today. So a vision of simplicity always meets with resistance. Often, skeptical stakeholders will try to block change.

It's tempting to deal with this by talking about the gain and avoiding talking about...the pain. But the pain is real, looms larger in people's imaginations, and comes before any gain. You must let people talk about it so they can come to terms with it. Do so, and they'll often talk themselves out of their fears.

Nevertheless, that conversation is complex. If you're talking to many stakeholders, the conversation often goes around in circles, so it helps to sketch out the three stages of change.

The first stage is the current, all-too-comfortable, situation. The question to ask is, Do you think the solution we have today will still be in use in a hundred years' time? Of course not. Change is inevitable. We simply need to know when it's advantageous to change. It helps to gather evidence to support the need for change (maybe you know your competitor is about to release an update).

The second stage is the pain. When you change something, some of your existing customers will complain, your colleagues will have to adjust, and performance will dip. The question is, How deep and how long will that dip be? That's hard to estimate, even with experience or user research on prototypes. The point is, you know the pain must come, so you can discuss how much you're prepared to bear.

If the change is long and complex, then you may want to run two products in parallel (many websites test changes with a subset of users before gradually releasing them to everyone).

The final stage is the gain. If your new designs are good, you'll see an uplift, which will eventually make up for the pain. Your user research can give you an indication of how big that uplift might be. Sometimes, it's enough to demonstrate that there will be an uplift. Other times, people will want more data, and you'll need large-scale research.

People usually convince themselves of the need for change. Make it their problem: Ask "What will you do about the inevitable change?" Let them talk through their fears, and encourage them to move through the timeline.

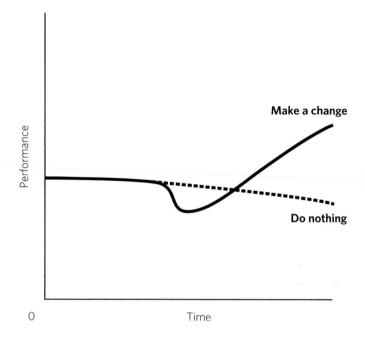

**Strategic changes
pay off over time.**

Vision and strategy

As I talk to designers, at every level, in every country I visit, in every type of organization, the same question comes up: How can I get my stakeholders to listen to me? The answer is that you must understand what's in it for them.

Your vision should tell you about your customers, their needs, and how you intend to solve those needs. But to complete the picture, you also need to link that vision back to the organization for which you're working. You need to understand your organization's strategy.

Strategy is a word I hear thrown around a lot by people who want to sound impressive but who don't always have a clear idea of what a strategy is.

The best definition I've come across is disarmingly straightforward. A *strategy* is a diagnosis of the current situation, leading to an insight into a superior solution and a long-term plan for winning.

According to Michael Porter, the author of *Competitive Strategy: Techniques for Analyzing Industries and Competitors*, there are three generic strategies for how a company might pursue competitive advantage in its chosen market. Companies can choose to be different from their competitors (and therefore command a higher price), to operate at a lower cost (and therefore derive greater margins, or more likely charge less), or to specialize by focusing on a particular market or need.

Simplicity looks different depending on which of those strategies your organization is pursuing. If you're differentiating, then you're simplifying to make something uniquely easy (like Stripe's online checkout process). If you're cutting costs, then you're simplifying to drive down costs (like Southwest Airlines). If you're focusing, then you're simplifying to optimize for a specific audience (like GoPro cameras for sports enthusiasts).

Porter felt that organizations should focus on one strategy, but increasingly companies are pursuing hybrid strategies (for instance, by differentiating themselves among other budget airlines). There's always a primary strategy ("be a budget airline") and a secondary one ("be the most enjoyable airline"). In other words, there's always a priority to the design choices you can make.

Whatever the strategy, the goal is to build a secure audience of customers to ensure survival (by generating revenue or ensuring relevance).

Understanding your organization's strategy helps you link your vision to the company's goals and tells you what kinds of simplicity to pursue.

Stripe differentiates its service from competitors by offering a uniquely simple online checkout.

The simple equation behind every business

Early in my design career, I thought my job was to come up with new ideas for companies. I quickly learned that companies didn't need more ideas. There were always ideas, and most often companies were overwhelmed by the number of ideas they had. My job, most often, was to help them pick the right ideas and make sure they were executed well.

Teams often capture those ideas on a "backlog" of changes to be made. But the problem remains which ones to pursue. Many organizations do this in a rather haphazard way. Some will pick out ideas that the team likes the sound of. Others will do whatever the next big customer is requesting. Many find themselves chasing the idea that the boss had in the bath the night before.

Here's a better approach, which I learned from Peter Merholz, author of *Org Design for Design Orgs: Building and Managing In-House Design Teams.*

Most companies are driven by an equation. Something like:

(number of cars sold) x (price of car) - (cost of overhead) = (profit)

You need to understand how simplifying the user experience could affect each of those elements. Will making the products simpler enable the company to sell more cars (for instance, because they'll be more desirable) or at a higher price (because they'll be seen as more sophisticated) or at a lower overhead (because the components will be less expensive)?

Next, you need to prioritize those changes. A good way to do this is to plot out how important each change is versus how feasible it is. If you ask people, they'll tell you that everything is important and anything is feasible. Instead, get them to divide up a fixed number of points (or Monopoly money or jelly beans) for importance and a fixed number of points for feasibility.

The changes that sit at the top-right corner of your graph are your priorities, and they are what your improvements need to address. Do that, and you'll have a prioritized list of changes.

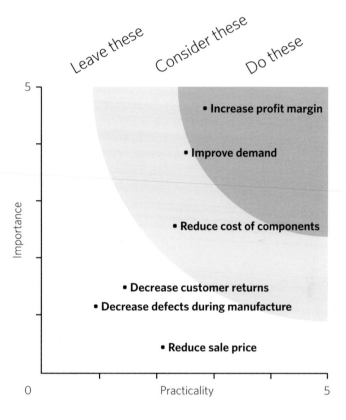

Leave these   Consider these   Do these

5

Importance

• Increase profit margin

• Improve demand

• Reduce cost of components

• Decrease customer returns
• Decrease defects during manufacture

• Reduce sale price

0   Practicality   5

Breaking free of "quick wins"

Prioritizing changes based on cost (or feasibility) alone leads to problems: The backlog of potential changes becomes choked with "quick wins." Focusing on "quick wins" stores up problems for teams.

One company I worked for ran a stock-keeping system that had been built in the late 1980s. It had never been intended to be put online, and to save money the developers had decided that stock records should use capital letters only. Cutting out lowercase letters meant data could be encoded more efficiently and halved the cost of storage.

Two decades later, this system was still in use, and when customers looked at product details online, they were in all caps. Ugly, difficult to read, and unprofessional. The company knew it was costing them sales and customers, but replacing this system was an expensive infrastructure project, so it kept on being put off in favor of "quick wins." Eventually the cost ran into millions upon millions of dollars. If you focus only on quick wins, it will kill you in the long term.

Breaking free of quick wins means doing two things.

First (as we saw above), it's important to look at "impact" as well as "feasibility" (or cost).

Second, you should separate changes into three types: quick wins (things that have a rapid impact), infrastructure wins (changes that have a deep impact or that make other changes more efficient), and strategic wins (things that bring your product into alignment with your strategy and vision).

A good product management approach seeks to divide time between all these things—to pay back in the short term and the long term. By separating changes like this, you acknowledge the need for different types of work to be done, and you compare like with like.

Just as smart investors seek to balance their portfolio across a range of investment types, good design teams try to balance their work across these three types of change.

**The best choice
isn't always the
low-hanging fruit.**

Small steps to big changes

Organizations often flip-flop between wanting to see big exciting changes and wanting to make safe, manageable changes. For design teams, this can be confusing and frustrating. If you're trying to come up with a sensible plan to simplify a design, you'll often find that some people think you've gone too far, while others don't think you've gone far enough.

It's worth remembering that organizations are made up of groups of people with different priorities. Talk to the board of directors, and their job is to take a long-term view. Boards often love big, flashy projects. Talk to an operations manager, and he or she is tasked with keeping things running smoothly. Operations managers like small, safe projects. No matter how big or small, organizations contain both types of people, so you'll always encounter conflicting priorities.

One way to cope with this is to stop focusing on the changes and start focusing on the users' problems instead. Say you have a user problem: Make it easy to apply for a credit card online.

The simple, safe solution might be to make a few changes to the online forms. The big, flashy solution might be to let users take some photographs of their banking documents and use some smart technology to fill in the forms for them.

Instead of seeing these as competing solutions, you can see them as steps toward a strategic goal of simplifying the application process.

The first step is the smallest change that signals to users and the team, "We care about this." It makes things better and gets everyone engaged with the problem. It's safe and it will generate a rapid return on investment. But the board will be...bored.

The next step is to learn how to deliver the flashy solution. So try out the technology in a limited way (for instance, scan in your driver's license to fill in a few basic details).

The last step is to deliver at scale—to go for that big flashy solution that the board wants.

By aligning your ideas to user problems, you can set out road maps for simplicity that get you from "a little better" to "wow" and help you to get buy-in from throughout the organization.

Some stakeholders want a safety-first approach.

Sweating the details

Research any simple experience and you'll be surprised at the amount of time design teams have spent focusing on tiny details. When colleagues of mine were designing an online supermarket, they spent weeks refining the way that shoppers added items to their basket. To them, the importance was obvious: This interaction happens dozens of times during each visit. Simplifying that one interaction could transform how simple it felt to buy groceries online.

Still, it's not always clear-cut why details matter. Stakeholders can object to the amount of time a team spends refining a design that seems good enough.

Elite sports teams know the importance of focusing on details. When he was performance director of the British Olympic Cycling Team, Dave Brailsford championed a system he called "aggregation of marginal gains."

"If you broke down everything you could think of that goes into riding a bike and then improved it by one percent, you will get a significant increase when you put them all together," Brailsford explained.

That philosophy led the team to change everything from their pillows (to improve sleep) to training in wind tunnels (to improve posture and aerodynamics) to getting surgeons to teach the team how to wash their hands (so they lost fewer days training due to illness). It also led them to become the most successful cycling team by far across successive Olympic Games.

It's a philosophy that other sports teams have taken up. When the Red Bull Formula One racing team won the World Championship in 2013, they redesigned their cars' wheel nuts so that pit crews could fit and tighten a wheel in a single motion. That led to the (then) fastest-ever recorded pit stop timed at 1.923 seconds to change all four wheels and send the car off again. Red Bull won the Constructors Championship at a canter.

If you're the only person on the track, then "good enough" is all you need. But most of us face significant competition. That's when marginal gains—the details—become important.

As Charles Eames said, "The details are not the details. They make the design."

The difference between the best and the rest is in the details.

Simplify this

Whenever I invite someone for a job interview as a designer, I ask them to show me how they'd take something that seems unnecessarily complicated and simplify it. There's no "right" answer—I'm interested in how they think and explain their ideas.

For a long time, I've been giving people the task of simplifying a remote control because most people have one at home and because, as we'll see, it presents some tricky problems.

Typically, a remote control has more than forty buttons; many have more than fifty. That seems excessive for a device that's used to play and pause movies.

When something is that complicated, there should be plenty of scope for simplifying it. But the task turns out to be harder than you'd imagine.

Try it now. You can refer to your own TV remote or use the template on the following page. You may find it helps to discuss the problem with a friend, but I wouldn't do this while they're trying to watch television.

Input select/text hold

On/off

Pause live TV

Widescreen mode

3D display mode

Music/video track ID

Rewind to start

Rewind

Fast forward

Forward to end

Record TV

Play recording or disk

Pause playback

Stop playback

Red/Green/Yellow/Blue multifunction buttons

iManual

Sync menu (link a compatible device to TV)

TV guide

Browse online movies

Program information/reveal text

Return/back

Home menu

Options

Four-way cursor

Select/confirm

Digital/analog TV mode

Exit

TV/radio mode

Numeric pad

Text (for services such as news)

Subtitles

Mute

Volume increase/decrease

Change program/scroll text

Audio mode (stereo/mono,
turn on/off audio descriptions)

The remote control

You can use the template on the opposite page for your remote control. The descriptions are the same as in the instruction manual for my own television, but I've added some explanation to a few of them. Most people would just have the icons on the remote control to go by.

Sometimes solving one problem leads to others. Try to think about how you'd use your version of the remote control and the ways in which it might feel simple or more complicated to use. Don't stick with the first design you come up with. It's always better to make three or four quick sketches than to grow too attached to one idea. Once you've done that, you can select a favorite and try to complete it.

I've been collecting people's designs for a while. If you'd like to see some of them and add your design to the list, visit simpleandusable.com.

Input select/text hold

On/off

Pause live TV

Widescreen mode

3D display mode

Music/video track ID

Rewind to start

Rewind

Fast forward

Forward to end

Record TV

Play recording or disk

Pause playback

Stop playback

Red/Green/Yellow/Blue multifunction buttons

iManual

Sync menu (link a compatible device to TV)

TV guide

Browse online movies

Program information/reveal text

Return/back

Home menu

Options

Four-way cursor control

Select/confirm

Digital/analog TV mode

Exit

TV/radio mode

Numeric pad

Text (for services such as news)

Subtitles

Mute

Volume increase/decrease

Change program/scroll text

Audio mode (stereo/mono,
turn on/off audio descriptions)

The four strategies

Over the years, I've seen many ingenious solutions to the problem of simplifying a TV remote, but I've found that they fall into four categories:

- Remove—get rid of all the unnecessary buttons until the device is stripped back to its essentials.

- Organize—arrange the buttons into groups that make more sense.

- Hide—hide all but the most important buttons behind a hatch so they don't distract users.

- Displace—create a very simple remote control with a few basic features and control the rest via a menu on the TV screen, voice controls, or gestures, displacing the complexity from the remote control to the TV.

Most people do a little of each, but usually they pick a primary strategy. Some use additional technology like touch-screen displays on the remote control or the ability to wave at the TV, but these are just forms of removing, organizing, hiding, or displacing.

As I've tried to simplify other devices and experiences, I've found that the same four strategies keep cropping up in one form or another. The strategies apply to both functionality and content. And the strategies apply whether you're looking at something large, like an entire website, or something small, like a single page.

Each of the strategies has its strengths and weaknesses, which I'll discuss in the following chapters. A big part of success comes in choosing the right strategy for the problem at hand.

Remove

Organize

Hide

Displace

Part 4

Remove

Remove

According to a 2002 study by Standish Group, 64 percent of software features are "never or rarely used." Look at your TV remote control and count the number of buttons that you've never touched. The same goes for almost any gadget or software you care to name. There are plenty of opportunities to simplify by removing.

Conventional wisdom says that more features mean more capability, which, in turn, means a more useful product. Conventional wisdom also says that products with more features will beat products with fewer features. But simpler products frequently displace their more complex rivals.

In the 1990s, Clayton Christensen, author of *The Innovator's Dilemma*, looked at why big companies with plenty of cash and dominant products were unable to sustain their success and discovered something surprising.

He'd assumed that technology moved on and the big companies just couldn't keep up. But it turned out they often led the way, thanks to their well-funded R&D departments. And it wasn't bad leadership—the big companies were run by smart people and backed by investors in the stock exchange. A lot of people thought they were making the right decisions.

It turned out the that technologies that overtook them tended to be worse. Sophisticated, integrated steel mills were overtaken by cheaper mini-mills. Beautiful home radios were overtaken by poor-quality transistor sets. Desktop computers were overtaken by smartphones and tablets.

The simpler products were cheaper to make (so they cost less) and easier to use (so they found a wider market).

Removing clutter allowed designers to focus on solving a few important problems really well. It also allowed users to focus on meeting their goals without distraction.

It's often easy to understand what's essential: A TV remote needs a way to switch it on and off, change channels, and set the volume. But does that mean you should cut everything? The French author and aviator Antoine de Saint-Exupéry said that "Perfection is finally attained not when there is no longer anything to add, but when there is no longer anything to take away." Knowing when you have reached that point is what makes removing such a difficult strategy.

The most obvious way
to simplify is to remove
what's unnecessary.

What not to cut

When you're up against a deadline, it can be tempting to drop hard-to-build features. People justify it to themselves by saying they're launching a Minimal Viable Product (MVP) and that they'll add the features later.

But an MVP should align with your vision and offer something of value. And adding features later only works if you know how you'll find the time and money to do that.

A few years ago, I worked on a website that was intended to help people conserve electricity. The big idea was to let people track their electricity usage online and see how small changes in their habits could lead to big savings.

When it came time to begin the design, the project manager decided this feature was too difficult to deliver and dropped it in favor of publishing some articles about saving electricity. When the site launched, there was nothing compelling or original about it and it failed to gain the intended audience.

This is a common pattern. A deadline approaches, a budget tightens, and features are cut. Frequently, the team focuses on delivering as many features as possible. Those that are big or tricky to deliver are cancelled. If someone objects strongly, they're told their feature will be pushed into "phase 2" or "phase 3."

What's left behind often adds up to an uninspiring product that's similar to a lot of existing, mediocre offerings.

This approach can tear the heart out of a project, and yet it's the standard approach to removing features and content, and one I've encountered far more than any other.

You can't avoid the process of removing features and content. Every team has limited resources, and every design project I've encountered has reached the point where features or content needed to be cut. It might be a product that had grown too big over the years or a new design that had to be reined in.

Don't wait for the unsympathetic, unsatisfactory process of cutting the most interesting features. Take charge of the design and ensure that you're focusing only on delivering features and content that add value to the user's experience.

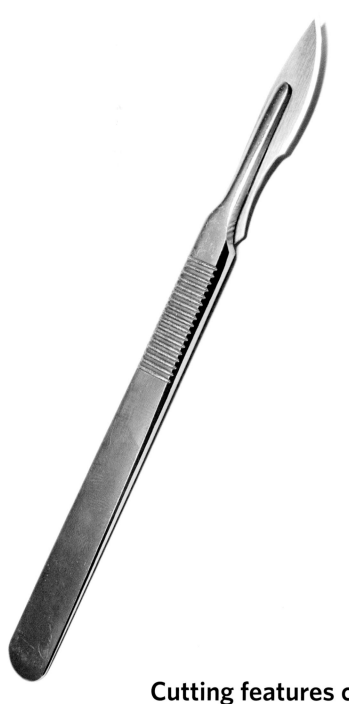

Cutting features can be a bloody process.

Find what's core

The core of the experience is the thing that matters most to users. Find that, and it becomes clear what to cut and what to keep.

At Telewest, Alan Colville was asked to design a new set-top box incorporating a "record" feature.

With tight resources, Telewest couldn't deliver everything on its wish list, but the company was paralyzed over what to drop. So Alan started user-testing competitors' products to see what mattered to customers.

To his surprise, he found that customers were frustrated by recording shows while they were watching television. Their favorite shows were often on at the same time and overlapped with each other. If they tried to record two TV shows, they couldn't watch a third. People complained that often they'd be recording two overlapping shows and wouldn't be able to change channels.

Overcoming this problem required adding a third TV tuner to the box—a major design change. But Alan's research showed that customers' frustration with this point was stronger than their interest in value-added features such as "red button" applications and interactive TV services, both of which had strong business support but weak customer need.

The research convinced the managers to switch their resources into the additional tuner. It quickly came to be seen as a competitive advantage and *Which?* (the UK equivalent of *Consumer Reports*) points to this flexibility as the box's major advantage.

When you're prioritizing features, remember that users value features that relate to their everyday experience of a product. That's subtly different from their most frequently used features. Begin by following the path set in your vision story. For a DVR, the ability to record and watch TV is closer to this everyday experience than pressing a red button to get mysterious "extras."

Users also value features that eliminate their frustrations effortlessly. When you're plotting your vision story, look for common frustrations and problems. Features that address these are your next priority. For a DVR, the ability to watch and record several shows at once turned out to be important enough to make it a priority.

**Customers preferred
basic improvements
to value-added extras.**

Kill lame features

It's often a good idea to get rid of poorly implemented features. While he was Head of Online at TUI Ski, David Jarvis recalls that one of the websites he managed had features that let users filter search results and create shortlists. He says:

> Neither was implemented particularly well. Although both filtering and shortlisting are features we thought should have been part of the functionality, and although we'd got something that was kind of working, we felt we were giving people a half-baked experience. We took the features off the UK site and our conversion rate went up.

One objection to removing half-baked features or content is that the time and effort that has gone into creating them will be wasted. No matter how poor the item, if it's been paid for, no one wants to get rid of what they have. In the words of Jack Moffett, "Broken gets fixed. Shoddy lasts forever."

Economists call this the "sunk costs fallacy." In reality, the cost of creating the feature can't be recovered, so the only way to judge the feature is on how much good it is doing and how much more it will cost to keep.

Features and content always place a mental load on users ("Do I look at this or not?") and always cost something to maintain (someone will have to keep the content up to date or make sure the feature still works).

In other words, features always have a price to you and your users. They should add value. If they're broken, deliver half of the answer, or duplicate functionality from elsewhere, then that value is diminished.

The question is never, "Why should we get rid of it?" It is always, "Why should we keep it?"

One reason to keep something: I'd take a poorly implemented safety feature over no safety feature. But *fixing* shoddy safety features should be a high priority.

Hanging on to features "because getting rid of them would be a waste" may be holding you back.

We tend to keep things, even when we know they're broken.

What if the user...?

If you've ever experienced design by committee, you know it can be impossible to argue that *anything* is unnecessary.

You start off with an idea of which features to kill, but, one by one, they are all justified with the words "but what if the user wants to...?" Sitting around a conference table, it's easy to imagine that, yes, a user might want to do just that. So the feature stays. By the time you get to the bottom of the list, you have most likely added a few more.

The "What if the user wants to...?" test allows any feature to get back into your product. If all a feature has to do is meet the "what if...?" test, then your plans will become choked with irrelevant junk. Like a traveler cramming his suitcase for every possible eventuality, you'll find yourself crushed beneath the weight of "what if...?"

It's fine to ask yourself "what if...?" when you mean "what if we solved the problem by...?" Dreaming up new ways of fixing things is one way to make your users' lives better.

What's *not* fine is using "what if...?" to dream up new problems or to guess at what's important to your users. Saying "what if the user wanted to...?" is a way of scaring people into imagining they have missed something. To cope with that fear, people are asked to divert time, effort, and money into adding features.

So "what if...?" can lead to fear that takes a powerful hold on meetings.

If you find yourself (or anyone else) saying, "What if the user needs to...?" there's only one answer: Go find out whether it's really important to your users. Ask, "How often do the people I'm designing for encounter this problem?" If the answer is "hardly ever," then drop the idea and move on.

Stop guessing "what if...?" and go find out what *is*.

**Avoid speculating
about what users
might or might not do.**

But our customers want it

Jürgen Schweizer of Cultured Code warns against adding features simply because customers ask for them:

> We get a lot of feature requests, but what our customers don't always realize is that if we went ahead and put an idea straight into the product, we'd probably break it. It would be too much or we'd have to move something important. So we try to resist adding new features.
>
> Instead, we try to reverse engineer the ideas—to figure out what problem the customer was having and to think about whether or not it's something we should try to solve in our software.

Features often involve trade-offs that customers aren't always aware of. Letting applications run in the background on your mobile phone sounds good—until you realize how quickly that can drain your battery and how annoying it can be to find out which apps are running and turn them off manually.

Adding features doesn't always make the user's experience simpler. Often it can lead to more frustration.

Sometimes you may be able to come up with an alternative solution that meets customers' real needs (such as letting them switch between mobile applications quickly). But don't be afraid to ignore requests to add more to your product.

Listen to your customers. But don't take their suggestions at face value.

Features that trigger errors

When I was working for an online bank, the manager in charge of savings accounts asked me to add a feature that would allow customers to divide their savings accounts into "pots" that they could name ("holiday," "gas bill," and so on). This would help customers to become more diligent savers by seeing what they were saving toward.

When I started to design the process, things quickly became complicated. For instance, when a customer added money into a savings account, they would need to add the money and then move it into a pot—two steps instead of one. When someone else added money into that account they might not know about the pots, so the money would need to go in another "general" pot in the account.

It became even more complex when transferring money from the account. The customer had to specify which pot the money should come from. And if the customer transferred too much money from that pot, they might be denied, even if there was enough money in the rest of the account.

When a feature is supposed to work one way but contains lots of exceptions to the rule, it sets off an alarm in my head. Those exceptions lead to confusion and errors.

So I went back to the problem we were trying to address: helping customers remember why they were saving.

I realized that if we allowed customers to name their savings accounts, we'd have a similar effect to naming pots. If customers wanted another pot, they could open another account. We could even start them off with two or three accounts and suggest names for them. Compared to the cost of implementing, explaining, and supporting the pot feature, naming accounts was quick and cheap to implement. And it was far easier for customers to understand.

If you design by trying to make a process work, you'll often find yourself drawn into creating features to handle exceptions, problems, and details. If you want to remove all this complexity, then step back, focus on the customers' goals, and ask yourself, "Is there another way to solve this problem?"

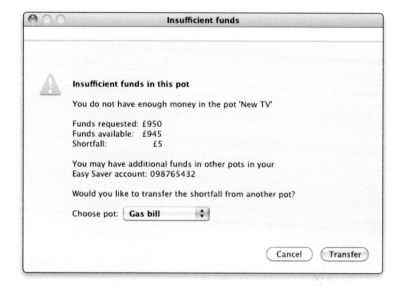

Insufficient funds

⚠ **Insufficient funds in this pot**

You do not have enough money in the pot 'New TV'

Funds requested: £950
Funds available: £945
Shortfall: £5

You may have additional funds in other pots in your
Easy Saver account: 098765432

Would you like to transfer the shortfall from another pot?

Choose pot: **Gas bill** ▲▼

Cancel   Transfer

When a small change leads to complex processes, it's time to step back and find another solution.

Errors

Even small errors can add to the load on users. Squishing them is a great route to simplifying a user experience.

A few years ago I was asked to design a current account display for an online bank. The bank wanted the service to match their brand values: friendly, approachable, and simple.

On the current account screen was a control that allowed the user to choose a bank statement. The user selected the month and year of the statement from two drop-down menus and clicked "Go." It seemed simple enough.

But the control could generate two possible error messages. If you selected a date in the future, an error message came up that said, in effect, that you'd been stupid. If you selected a date that was over a year old, you were told to try again, since the bank only kept statements for a year. A person in a hurry could easily make either mistake, and neither error message was particularly friendly, approachable, or simple.

The problem was that the user was being asked to enter a date, when really he needed to choose from the last twelve bank statements. So I replaced the two date controls with a single drop-down list of the available bank statements.

With the redesigned control, users could only select from what was available, so there were no error messages to design. This made the system simpler to maintain, too.

Whenever a user has to correct an error, it breaks his concentration and makes the experience feel more complex. Designers often try to prevent errors by interrupting the user ("Are you sure you want to do that?"), but in a way this approach is worse because it interrupts everyone, whether they've made a mistake or not.

When you're trying to simplify an experience, looking for places where error messages are needed, or checking the error logs for common error messages, is a critical step.

Removing sources of errors is an important way to simplify an experience.

If you forget to change the year, you can accidentally request next month's bank statement. The redesigned interface simply lists the available bank statements.

Before

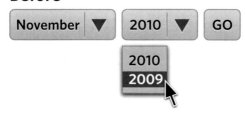

After

When features don't matter

If you're trying to make an appealing product, getting rid of features seems risky but has long-term benefits.

In 2006, three researchers—Roland T. Rust, Debora Viana Thompson, and Rebecca W. Hamilton—conducted an experiment to see whether features or usability mattered most to customers.

They divided participants into two groups and asked them to choose between two digital video players—one with seven features, the other with twenty-one features. Participants from the first group were only allowed to read about the products before they made their choice. The second group got a chance to use one of the products (either the high-feature model or the low-feature one) before making their choice.

Two-thirds of participants in the "no use" group chose the high-feature model. But only 44 percent of participants who used the high-feature model went on to choose it—and they were less confident that they had made the right choice.

Their conclusion: Feature lists sell so long as customers don't get a chance to use the product. But once consumers have used a product, their preferences change. Suddenly usability matters very much.

Today, word of mouth, user reviews, personal recommendations, and product trials are becoming more important than mass advertising. Customers find out about products from other users—people who've learned to value usability. There's a strong argument for cutting features, rather than piling them on.

Overburdening your product with features is likely to decrease mainstream users' satisfaction and hurt sales in the long run.

**In the long run,
adding features is
a losing strategy.**

Will it hurt?

Once a feature has been released, someone, somewhere, will eventually use it. If users like it, they will change their behavior to take advantage of it. People become addicted to their favorite features, and they will be irritated when one is taken away, no matter how trivial the change.

But some addictions are easier to break than others. What matters most to your users is this: Is your design best at solving their big problems? If it is, they will stick with you, even when they're inconvenienced by your changes.

Judging how much the removal of a feature will affect users is a delicate business. Simply asking people, "Would you like us to remove this feature?" always delivers a resounding "No!" No one likes the thought of getting less. Even people who never have and never will use the feature will want to keep it. The *idea* of features is often more appealing than the reality.

Instead, begin by assessing how close it is to the users' core goal.

If you're designing a mobile application to help salespeople organize their leads, removing a feature that changes the background color won't hurt; it's not a core task.

But if the feature is closer to the core of the application, things are a little harder.

Watching people use mock-ups is the best way to find out what really matters and to understand how they will respond.

Trying to please all users all the time is an impossible task. Aim to delight your target audience for their core tasks and hope to please them for the secondary tasks.

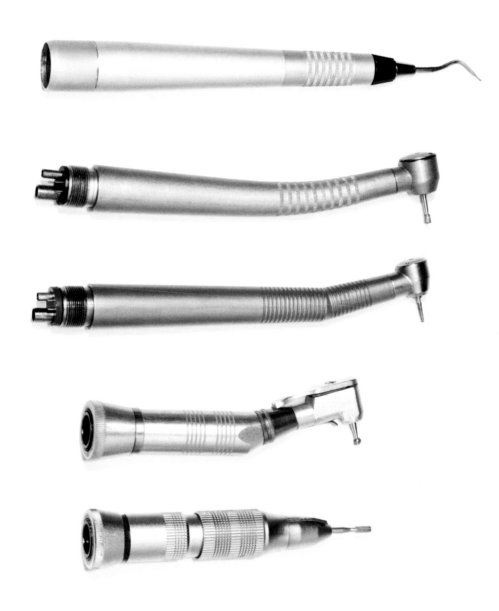

Some people find it hurts more than others.

Prioritizing features

When you're trying to figure out which features to keep and which to remove, follow these principles:

- Identify the users' goals and set them in order of priority. For the TV remote control, a main goal is to find a channel to watch and set the volume (I'd include "subtitles" in this); a secondary goal would be to switch to a different video source like a streaming video player; a less important goal would be to adjust the picture.

- Focus on solutions that completely meet users' high-priority goals. Only then move on to the lower-priority goals. Don't botch what's core in favor of delivering more stuff.

- Identify things that are common sources of anxiety or stress and prioritize features that alleviate that stress effortlessly. For instance, interruptions (such as the telephone) are a common frustration when watching TV. The pause button on a DVD remote control is a way of minimizing that frustration.

- Identify the controls that satisfy mainstream users' needs (in general, those are the ones that allow "good enough" control) and the controls for experts (in general, precision control and customization). Set aside the "precision" controls or replace them with "good enough" alternatives.

- Remove duplicate ways of performing a task. For instance, the TV remote control in this book has two different controls for scrolling text on the screen. Pick one.

And finally, don't be tempted to judge the value of your product by the number of features. Instead, consider how well it meets users' high-priority goals. In other words: Watch people using prototypes.

**Prioritize features
that satisfy
mainstreamers' needs
with minimal effort.**

Load

People have a limited capacity to process information, learn procedures, and remember details. And in the real world, they're under far more pressure from interruptions and deadlines than in a user-testing lab, which limits their capacity even more.

Small details in an interface can add to the load on the user and slow them down like speed bumps and potholes on a road.

When The Co-operative Bank asked my business partner, Richard Caddick, to increase the number of people clicking through their home page, he set out to reduce the load on people visiting the page.

- He removed text that was not being looked at, such as the tagline underneath the bank's name.

- He simplified the layout, removing a vertical column on the right side of the page so it was easier to see which items were important and which were low-priority.

- He eliminated duplicate links, such as the "Tell me about..." drop-down menu, cutting the number of clickable items by about 20 percent.

- He limited the number of styles used for buttons and links to make it easier to distinguish what was clickable and what was not.

- He reduced the number of promotional slots so there were fewer distractions for customers who knew where they were going.

- He cut down the visual clutter by removing distracting elements such as lines that were used to divide content and a horizontal yellow bar across the page.

This small project took just a few weeks to complete, but it resulted in a significant boost to the number of visitors clicking through the home page and going on to complete application forms.

Removing options, content, and distractions lightens the load on users so they can focus on getting the job done. Removing visual distractions helps them process what they're seeing faster and more reliably. Remove distracting details.

Before

After

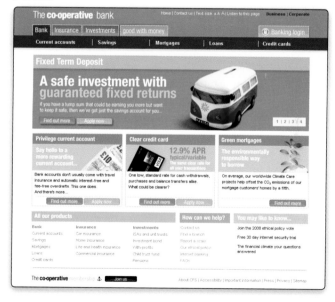

Decisions

We often focus on giving users as many choices as possible. But choice can easily overwhelm users.

In 2000, Dr. Sheena S. Iyengar and Dr. Mark R. Lepper set up a tasting booth at Draeger's Market in Menlo Park, California. Hundreds of people walked past the booth each day. One weekend, they put out a selection of twenty-four varieties of jams; on another they set out six. The wider selection performed badly. Only 2 percent of passersby bought the jam. When there were fewer options, 12 percent of passersby purchased the jam.

Iyengar and Lepper repeated similar experiments in a number of settings, and found that people were more likely to make a purchase when given a handful of choices than when they were overwhelmed with dozens of options.

They also found that people who were given a limited choice were more satisfied with their selection than those who'd been given more options.

Offering people a choice gives them a sense of control, and people prefer some choice to no choice. But when that choice exceeds a handful of options it becomes a burden, especially when the options are similar.

You can see something similar at work in people's attitudes toward technology. Most people are anxious when faced with a massive array of options and buttons. Every time they pick up a complex gadget, there's a nagging sense that they don't fully understand it, and that a slip of a finger could easily make things go wrong. People can easily distrust choice.

When you're next looking at a long feature list, a webpage with dozens of links, or a computer menu that's full of choices, it's worth remembering how easily this choice can undermine your design.

Users are happier when their choices are limited.

Distractions

User interfaces are full of irritating distractions. These can turn even simple tasks, such a reading a body of text, into a chore.

Hyperlinks within an article may seem like helpful extras, but each link says, "Why not stop what you're doing and look at this instead?" They break into the user's consciousness and undermine her concentration. Researcher Erping Zhu has found that increasing the number of hyperlinks within a document lowers readers' comprehension—even if the reader doesn't follow the link.

The right column of a webpage often is often set aside for even more distracting links, ads, related content, and extra data. This stuff is usually brightly colored and animated to draw the user's attention away from the main focus of the page.

Users may well click the links, but if their journey ends in confusion, listlessness, or irritation, the distraction has been counterproductive. The best place for these extras is at the end of a page where the user has finished reading. If users aren't reading that far, then it's a sign that the article itself needs work.

Our devices, too, are constantly chirping for attention, sending us alerts demanding attention. We have little control over when they happen, and dealing with them can feel like a game of Whac-A-Mole. They feel stressful, not simple. Sadly, the constant interruptions and updates have an addictive quality. Our devices are alive, and maybe the next alert is important. Replacing distractions with simplicity is an important mission.

Human beings don't interrupt each other like that. If I have a message for you but I can see that you're busy, then I'll wait until you take a break. If the message is really important ("Your car is being towed") and your task is not (you're reading the sports news), then I'll interrupt. Humans make social judgements about when to interrupt. We recognize the value of focus.

Today, my social media apps don't distinguish by default between important updates and unimportant ones. Yet to you and me, the distinction is obvious. The reason, of course, is that the app owners are not concerned with our peace of mind. They want to interrupt us and get our attention at any opportunity.

If you're designing simple experiences, your job is to remove distractions and let the user focus.

Before

After

Smart defaults

You can reduce the number of decisions that a user has to make by choosing smart defaults.

Many car manufacturers' websites allow you to compare the model you're viewing with similar models. Click on the "compare" feature and you're asked to add two or three other models into a comparison chart. Lexus's European website doesn't do this. It prefills the table with the car you're researching and the two closest models. Often, the models it chooses are exactly the ones you need to make a useful comparison.

Some people may have to change the default selections, but they're no worse off than if the table were left blank. Overall, Lexus is saving time for its customers.

Smart defaults are ones that suit the largest possible number of people. Customer data, such as log files, provides a wealth of useful information for smart defaults.

- Popular documents ("Top news stories")
- Similar items ("Customers like you looked at...")
- Personal information ("Auto-fill the form with your address")
- Common choices (Putting "USA" at the top of an alphabetical list of countries because most of your customers come from there)

It's worth remembering that when a customer returns to a website or an application, he frequently wants to pick up where he left off.

- Recently saved documents ("Open hello-world.doc")
- Resume a process ("Continue your game from level 3")

A complaint I hear frequently from users of travel websites is how tedious it is to re-enter the same information every time they visit. Imagine how much simpler it would be if travel sites remembered the routes you typically fly or the hotels you usually visit.

Truly smart defaults "watch" what you (or people like you) do and look for patterns of behavior to anticipate your next move. The suggested replies in many messaging apps are good examples of this.

Defaults are a powerful way of saving users time, effort, and thought, and a great way to remove speed bumps from your design.

Lexus's default comparisons are useful for most customers, most of the time.

Options and preferences

When you're looking for something to remove, options and preferences are a good place to begin.

In general, options help users to customize their setup. This is classic expert behavior—experts want to get under their car and tinker with it; mainstreamers want to get in and drive.

I've found that options and preferences generally creep into designs when the design team isn't sure what to do. Maybe there are two possibilities for navigating a website: breadcrumb links or drop-down menus. Both look good, so both go in. That way the user has a choice.

This sounds like it would be helpful, but should users be wasting their time figuring out which navigation technique is most convenient? That task is so far removed from a vision of a simple product that it never appears. Let's go back to the Jennifer Lawrence story from Part 2 for a minute. Imagine: You hand your camera to a friend who then determines which of the three available grip positions and shutter buttons is best. Your friend would be wasting precious time, and you'd probably miss your chance to take the video.

Simple user experiences don't force the user to make these kinds of choices. It's the responsibility of the design team to do that. The best way to decide is to try it out on some users. And if there's no clear winner, and no dangerous pitfalls, then there's no "wrong" design. Choose which one to implement and move on.

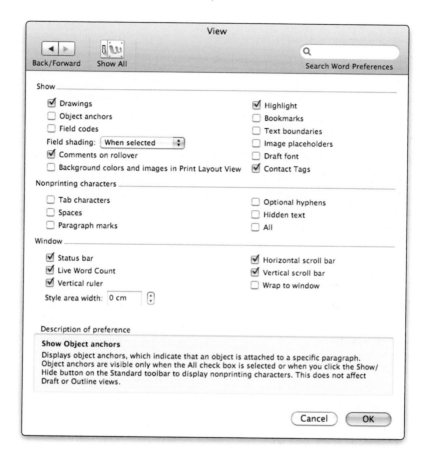

Mainstream users don't like the burden of setting options and preferences.

When one option is too many

Sometimes, even one option is too many. A while back I watched a user test of the special offers section of a travel website. We asked participants to find and book a holiday. They easily found holidays that they wanted and announced they had made a decision. But next to the booking button was a link to "look for more deals." This proved irresistible. Every time a participant came close to booking, she clicked that link. No one booked a holiday.

We'd assumed that the link would help people who weren't quite sure. Instead, it undermined the confidence of everyone who came close.

When you're offering a choice to your users, think very carefully about whether you're overwhelming them with options or undermining their confidence in their decision.

Look at the design of the online checkout on any big site like Amazon or Best Buy. The checkout is where users have to make a choice: buy or bail. The retailers know that any doubt will undermine users' willingness to complete the transaction. So in the checkout, retailers remove navigation links that are normally at the top and bottom of every other page.

I doubt most customers are aware that this happens; when they get to those pages, they're too busy filling in forms. But retailers know that if they put those links back, customers will click on them and the sale will be lost. If this seems somehow underhanded, consider whether it is in the customers' interest to waste time by constantly dithering between website and checkout.

Remember, mainstreamers want "good enough, quickly;" experts want "perfect in as long as it takes." If you're designing the kind of simple experience that mainstreamers love, then ask yourself if the options you're giving them will sacrifice speed and simplicity for perfection. If the answer is "yes," remove the options.

What you mean

Browse other choices   Buy now

What they see

Keep my options open   No turning back
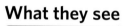

Visual clutter

Removing visual clutter means people have to process less information and can concentrate on what's important on the page. I've noticed that users describe interfaces they like as "clean," meaning free from clutter.

The designer Edward Tufte talks about needing to make the "data-ink ratio" as high as possible. In other words, ink (or pixels) shouldn't be wasted on anything that isn't content or in repeating content. So he removes the gridlines on graphs, leaving just the axes and the zigzag line of the graph itself. He reasons that the gridlines distract the reader from the important data: the shape of the graph.

The process for removing clutter is simple. Look at each element in the design and ask why it is needed. Is it critical information or there for support? Try to remove it from the design. If the design no longer works, replace the element.

Here are some good ways to limit visual clutter:

- Use white space or subtle background tints to divide up the page rather than lines. Why? Because lines sit in the foreground, so you pay more attention to them than tints or white space that sit in the background.

- Use the minimum possible emphasis. Don't make something bold, large, and red, if simply making it bold will do.

- Avoid thick dark lines where fine, light lines will do.

- Limit the levels of information. If you have more than two or three levels of information on a page, you may be confusing the user. For instance, limit the number, sizes, and weights of fonts. Try to keep to just two or three levels in total—e.g., a headline, subheading, and body text.

- Limit the variation in sizes of elements. For instance, if you're designing an online newspaper, you might have a large block of text for the main story and five smaller blocks of text for secondary stories, rather than six blocks of text in different sizes.

- Limit the variation in shapes of elements. Stick to one button style rather than using three or four different ones.

Q1 FINANCIAL FORECAST AND PERFORMANCE						
	JANUARY	JANUARY	FEBRUARY	FEBRUARY	MARCH	MARCH
	Forecast	Actual	Forecast	Actual	Forecast	Actual
INCOME						
Direct Sales	US$95,000.00	US$93,250.00	US$105,000.00	US$179,260.00	US$125,000.00	US$103,551.00
Interest Earned	US$750.00	US$731.00	US$800.00	US$759.00	US$900.00	US$787.00
Service Charges	US$11,400.00	US$12,075.00	US$12,600.00	US$12,397.00	US$15,000.00	US$15,010.00
Variation Fees	US$4,750.00	US$6,200.00	US$5,250.00	US$2,066.00	US$6,250.00	US$7,064.00
Other	US$12,000.00	US$1,976.00	US$7,000.00	US$3,291.00	US$9,000.00	US$4,884.00
TOTAL	US$123,900.00	US$114,232.00	US$130,650.00	US$197,773.00	US$156,150.00	US$131,296.00
EXPENSES	Forecast	Actual	Forecast	Actual	Forecast	Actual
Travel	US$2,500.00	US$1,137.00	US$2,500.00	US$4,320.00	US$2,500.00	US$2,403.00
Subsistence	US$1,000.00	US$576.00	US$1,000.00	US$1,433.00	US$1,000.00	US$998.00
Consumables	US$850.00	US$931.00	US$850.00	US$802.00	US$850.00	US$736.00
Hardware	US$3,000.00	US$2,944.00	US$9,000.00	US$8,975.00	US$3,000.00	US$2,780.00
Software	US$1,000.00	US$895.00	US$4,000.00	US$3,890.00	US$1,000.00	US$895.00
Comms	US$2,500.00	US$2,301.00	US$2,500.00	US$2,308.00	US$2,500.00	US$2,442.00
TOTAL	US$10,850.00	US$8,784.00	US$19,850.00	US$21,728.00	US$10,850.00	US$10,254.00
NET TOTAL	US$113,050.00	US$105,448.00	US$110,800.00	US$176,045.00	US$145,300.00	US$121,042.00
PERF		-US$7,602.00		US$65,245.00		-US$24,258.00

Q1 Summary (US$)

	January		February		March	
	Forecast	Actual	Forecast	Actual	Forecast	Actual
INCOME						
Direct Sales	95,000	93,250	105,000	179,260	125,000	103,551
Interest Earned	750	731	800	759	900	787
Service Charges	11,400	12,075	12,600	12,397	15,000	15,010
Variation Fees	4,750	6,200	5,250	2,066	6,250	7,064
Other	12,000	1,976	7,000	3,291	9,000	4,884
TOTAL	123,900	114,232	130,650	197,773	156,150	131,296
EXPENSES						
Travel	2,500	1,137	2,500	4,320	2,500	2,403
Subsistence	1,000	576	1,000	1,433	1,000	998
Consumables	850	931	850	802	850	736
Hardware	3,000	2,944	9,000	8,975	3,000	2,780
Software	1,000	895	4,000	3,890	1,000	895
Comms	2,500	2,301	2,500	2,308	2,500	2,442
TOTAL	10,850	8,784	19,850	21,728	10,850	10,254
NET TOTAL	113,050	105,448	110,800	176,045	145,300	121,042
PERFORMANCE		-7,602		65,245		-24,258

You'll be surprised how much clutter you can remove from a page.

Removing words

Why are so many webpages clogged with words that no one will ever read? Perhaps it's because, unlike paper, web pages can always accommodate more text, so it costs nothing to add another paragraph or two. Or three.

The extra text is often wasted. Users don't slavishly read every word. Their eyes skim over pages, picking out the odd keyword or sentence.

Getting rid of text has three benefits:

- It makes what's important stand out.
- It reduces the effort it takes to interpret a screen.
- It makes people more confident that they've understood what's there.

When you're hunting for text to cut, be aware of some common hiding places:

Skip the introductions. Often the opening text on home pages and in articles says nothing at all ("Welcome to our web site. We hope you'll enjoy..."). It doesn't sound chatty or inviting; it just leaves the reader wondering where the author is heading. Cut the intros and start with a bang.

Delete unnecessary instructions. These are frequently redundant and can be cut completely. Delete text like "Fill in the fields in this form and press Submit to send your application to us." The page title ("Application Form") and the contents of the page (a form) are enough to signal the user what to do.

Simplify explanations. Sometimes links have descriptions under them. These can be useful, for instance, when one audience expects the link to be called one thing and another expects it to be called something else. But often, explanations are another source of redundant text. Replace "Product Finder: Answer some simple questions and we'll find the right product for you" with "Product Finder," and you'll save twelve words from a total of fourteen.

Use descriptive links. Links called "Click Here" or "More" sometimes appear under the headlines that describe exactly where they go. Simplify the page by using the headline itself as the link.

"Get rid of half the words on each page, then get rid of half of what's left."
—*Steve Krug's Third Law of Usability from Don't Make Me Think!*
A Common Sense Approach to Web Usability

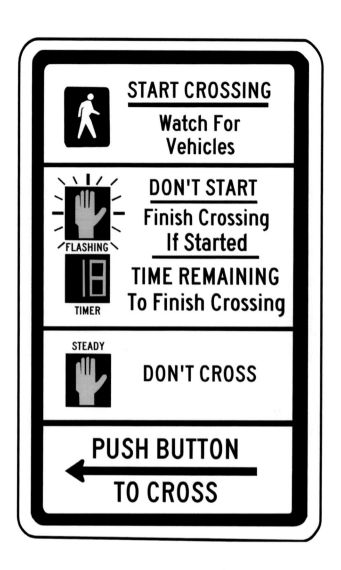

Avoid unnecessary instructions.

Simplifying sentences

Almost any sentence can be simplified, and almost any text can be cut. In *Revising Prose*, Richard Lanham offers a simple method to turn long-winded writing into short, crisp sentences.

- Circle the prepositions (of, in, for, onto, into, about). They drain the action from a sentence, so try to eliminate them.

- Circle the "is" verb forms ("is taking time") and replace as many as you can ("takes time").

- Convert passive voice ("time is needed for this project") into active voice ("this project needs time").

- Cut out slow starts ("One can easily see that...") and get to the point.

- Eliminate redundancies. Don't say "on a daily basis" when "daily" means the same thing.

These rules make text clearer, more persuasive, and shorter.

For example:

- Please note that although Chrome is supported for both Mac and Windows operating systems, it is recommended that all users of this site switch to the most up-to-date version of the Firefox web browser for the best possible results. (41 words)

Simplified version:

- For best results, use the latest version of Firefox. Chrome for Mac and Windows is also supported. (17 words)

Use Lanham's rules to remove the words that pad your sentences.

volkswagen.co.uk/efficiency

~~Our~~ BlueMotion ~~range combines lighter materials, enhanced aerodynamics, economical engines and tyres that create less friction, which~~ saves you ~~fuel and can reduce your tax, which means you will have more~~ money.

BLUEMOTION
— TECHNOLOGIES —

Another example of ~~Volkswagen~~ efficiency.

Das Auto.

DDB UK's advertisement for Volkswagen edits itself down to the bare essentials.

Conversation

An interaction is a dialogue between the user and the device. Listen to a conversation between two people and you'll notice something interesting that you can apply to the way that people interact with devices, software, and services: People adjust conversations to suit the situation.

When people are under time pressure, then *both sides* in the conversation recognize this and tend to cut things down to the bare minimum. Less pressure, and the conversation tends to be more detailed and open. Each side is looking out for the other, making sure they don't waste time.

When we deal with computers, they often have no way of telling the time pressure we're under. As a result, many interactions seem detailed, complex, and tedious: You want to make the booking online, and the computer is trying to get you to sign up to the loyalty scheme, too. What seems simple to a group of designers sitting around a table feels complex and frustrating to a user under time pressure.

Listen carefully to conversations between people, and you'll notice that things are a little more complicated than just taking time pressure into account. They also account for the risk (and consequences) of error. If one person thinks the other will run into problems, they'll slow things down—even if the other person is in a hurry. Often, they'll flag this to the other person ("I know you're in a hurry, but this is important...").

Finally, you'll notice that sometimes people skip details because they know (or assume) that the other person knows them already. Someone giving directions to a local may say: "The book store is right next to Starbucks," whereas they'd give turn-by-turn directions to a stranger.

Those three factors—time pressure, risk and consequences of error, and shared knowledge—are all factors that people use to simplify conversations. When you're designing an interaction, you can use those things to tune a conversation to feel simple according to your specific user's circumstances.

In conversation, people cut down their utterances
to take account of time pressure, shared knowledge,
and risk of error.

How do I get to the station?

Set out along Main Street and go ahead for 50 yards. At the junction with Spring Street bear left along Main Street for 240 yards. At the traffic lights go straight ahead for 70 yards. Take the ramp for the train station marked 'other traffic' and in 25 yards you'll be at the station entrance.

Uh, I'm in a hurry...

Follow Main Street for about 400 yards.

Cutting time

There's always a cost to using any software, device, or service, even if there's no monetary charge. Time.

Cutting features and content saves time because there are fewer decisions, fewer buttons to press, less thinking, and less reading to be done. It can also save time by making the service simpler and more efficient to build so that it is speedy to run. There's a reason people are willing to pay more for more powerful versions of the same computer: They save us time.

There's immediate value in improving performance. Most product owners don't look that deep. They spend budget on the features on the surface while the infrastructure underneath creaks under the strain. When things become critical, the job of fixing it is complex and expensive.

Many people assume that the more time people spend using their app or device, the more they liked it. My experience is the opposite. Whenever I've cut the amount of time it took people to choose a car, buy a movie, or read some information, the result has been more revenue, happier customers, and more users.

Watch carefully while someone is using a piece of software, and you'll be struck by how much time they spend doing nothing—staring at the screen, frowning, and hesitating. They're thinking, and thinking is effortful and hard. Sometimes it's useful to pause and reflect, but often it's a chore. Save users' mental energy for the things that really matter to them.

Time is, at best, an indirect measure of success. We assume that more time means users were satisfied or successful in their task, but the connection isn't that simple. If people are confident, they'll make decisions faster. If people are interacting with each other, they'll make use of a social media site. If it takes too much time and effort to do those things, they'll give up.

Of course, counting features and content is itself an indirect measure of the time it will take to complete a task. The thinking time is the critical part. So don't rely on lists or designs. The only way to know for sure is to test with users. But you can be sure that saving time will matter to them.

**Focus on helping
people to achieve
their goals quickly.**

Removing too much

In the Apple Store in Tokyo you'll find a remarkable glass elevator, finished in Apple's trademark brushed aluminum. What makes this elevator unlike almost any other in the world is that there are no buttons: none to call the elevator, and none inside. The lift shuttles between the four floors of the store, stopping at each one it passes.

Apple has reduced the elevator to its core: a platform for taking people between levels. But instead of feeling simple, it feels wrong. The elevator leaves you feeling unsettled, frustrated, and anxious. Will it stop at the floor I want? Why is it stopping when no one is getting on or off?

Apple has removed a crucial ingredient: control.

Without the sense of control (calling and directing the elevator) or the sense that a visible person is in control (the guy in front who just pushed a button for your floor) and the feedback that it's working (the button that illuminates when you push it), all you can do is hand yourself over to the machine and hope.

In the buttonless elevator, people waste time and attention worrying. Removing all control doesn't simplify the experience, it complicates it.

I've come across the same problem trying to get information from flight maps on airplane video screens. They switch from world map to local map to flight data agonizingly slowly. Not having any control makes that wait seem even longer.

People need to feel in control. They prefer to be pilots rather than passengers. When they're at the mercy of chance or hidden forces, they become so anxious that they invent superstitious behaviors that help them regain a sense of being in charge, like avoiding the cracks in the pavement or wearing a "lucky" shirt.

The trick is to give people control over outcomes, in other words, enough control to stop them from worrying that their basic needs won't be met, but not so much that they're wasting time making choices they don't need to make. (How fast should the elevator travel? How long should the doors stay open?)

No buttons in this elevator. But people prefer to be pilots, not passengers.

You can do it

Can a team within a large organization create a radical website design and convince stakeholders to remove content and features?

"The old home page was a billboard," says Fran Dattilo, the project manager for Marriott's home page redesign. "Everyone said, 'It's too cluttered. You've got to change it,' and everyone thought that their stuff had to stay on the new home page."

Marriott's user testing said the home page was a members-only club. It worked fine for regular customers, but newbies got lost and confused.

The home page redesign had to be flexible, but the user experience team discovered they'd created a monster that had grown out of control. They set about creating a design that was deliberately inflexible.

There were fewer content areas and only one featured item—the top item on a fan of cards. This slashed the number of links on the home page from 77 to 43, a big reduction in clutter.

To convince the company, the team gathered evidence. The new home page was the most tested design Marriott had ever launched, backed up with data from the live site. "We went back to our main stakeholders, and we could tell them that this link only got 500 clicks a year and that the new design worked in China as well as the U.S."

Even so, launch was stressful, recalls Mariana Cavalcanti, Marriott's Director of User Experience. "We came in at 3:30 in the morning to watch it go live. We had prepared the company for a 10 to 15 percent dip in bookings at first—that was important. But there was no dip. Satisfaction scores did fall—our regular users didn't see any need for change. But four months later satisfaction was above our previous levels. We still see a lot of comments on message boards comparing us to similar brands. We've made them look ugly."

Simple design is often said to be the work of a single visionary designer, a "ruthless" or "uncompromising" innovator. But most of us work in organizations where there's a lot of political give-and-take. Marriott shows you can simplify with a shared vision, a focus on the mainstream user, and a thoroughly researched design.

Before

After

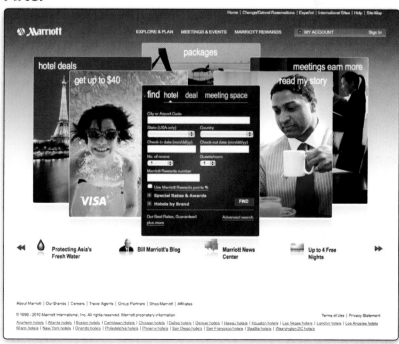

Focus

The "remove" strategy is about removing distractions to bring focus to your project:

- Focus on what's valuable to users. This means concentrating on features that deliver the users' core experience. It also means delivering features that eliminate users' frustrations and ease their sense of anxiety.

- Focus your resources on delivering value by removing lame features, irrelevant extras, and bribes.

- Focus on meeting users' goals. Agonizing over the process will get you bogged down in detail.

- Remove the distractions of tiny speed bumps that add to the load on the user: error messages, irrelevant text, unnecessary choices, and visual clutter.

With patience and the data to back you up, you can bring focus to most projects. If your problem is political, you can overcome it by building on small successes or by using evidence from tests. If your problem is out-of-date technology or incompatible systems, these too can change (slowly) over time. However, there are a couple of exceptions.

Sometimes there is an unavoidable legal requirement to include particular wording or information. Financial services and medical regulations often require that specific wording is used, not because it makes sense to the public but because it makes sense to lawmakers. Laws can be changed, too. David Sless in Australia has had some success in getting lawmakers to focus on whether consumers understand labels, rather than requiring long and confusing instructions.

Sometimes you can't remove because your design is part of a larger system. That's the case with the TV remote. For instance, many interactive services (like sports scores or financial news) are triggered by the red button on the remote. If you removed it, you would risk breaking the user experience for anyone who wanted to use those services.

While you can't change the rest of the world, there are always other ways of simplifying that you can adopt today.

**Removing clutter
helps users focus on
what's important.**

Part 5

Organize

Organize

Organizing is a great strategy for simplifying. In the case of the TV remote control, it's probably the solution I've seen most often. It's usually an inexpensive solution—changing the layout and labeling the buttons on the TV remote control costs less and demands fewer tough decisions than, say, removing buttons.

There are plenty of options open to you in organizing an interface—size, color, position, shape, hierarchy. But those choices need to be employed with restraint. Some of the TV remote controls I've seen over the years have had so many colored buttons they looked as though they were made from Skittles.

If you want to organize for simplicity, it's important to emphasize just one or two important things. Simple organization doesn't draw attention to itself; it lets users focus on what they're doing.

The best TV remote control designs emphasize the starting point (the on/off switch) and the most important controls (the channel and volume controls).

Your car is a great example of this. Sit in the driver's seat, and of the dozens of controls, it's the steering wheel, accelerator, and brake with which you naturally come into contact. It's a design that's lasted for decades. If design is like a conversation, then openings are always the most difficult part. Good design knows just how to say, "Hello, let's start here."

Organizing is often the quickest way to make things simpler.

Chunking

One way to make the blocks of buttons on the TV remote control more manageable is to break them into chunks.

Chunking is used throughout user interface design. Microsoft Word has hundreds of features. To make them manageable, they are divided up into around nine menus. Each of those has a couple dozen commands—still too many to take in at a glance, so they're divided into chunks again. Click a menu item, and it'll often take you to a dialog box where even more features are available. The daunting list of features is grouped into manageable chunks within a hierarchy.

The classic advice here is to break down items into groups of "seven plus or minus two." In theory, this corresponds to the number of items your brain can hold in short-term memory. If you read a list of ten items, you'll likely have forgotten one of them by the time you get to the end.

Many psychologists now believe short-term memory may be smaller—perhaps retaining just four items. But the "seven plus or minus two" rule remains, because it works. It seems to be a number that people can cope with. When I ask users to sort items into groups as they see fit, they tend to come up with around half a dozen categories.

There's no reason you can't divide the user's options into fewer chunks. I would always use as few chunks as feels simple to your mainstream user—fewer chunks mean fewer choices and less load on the user.

You don't always need to chunk. If your user needs to find an item in a long alphabetical list or timeline, there's no point in breaking up the list into half a dozen bits. Marking out letters of the alphabet or months of the year can help users quickly scroll to approximately the right place, but chunking is most effective when users have to evaluate several possibilities rather than locate an item on a continuous index or scale.

Organize into bite-size chunks.

Organizing for behavior

The first question a user will ask is, "What can I do with this?" so your first point of organization is to understand users' behavior: what they want to do and in what order they want to do it.

An online supermarket requires users to find the items they want to buy, add them to a shopping basket, schedule a delivery, and pay for the goods. Those are the main chunks into which the site should be divided.

People expect to begin their shopping by choosing groceries. This is also the most time-consuming part of the task, so it should be the most prominent.

People usually expect to do things in a particular sequence. It's disorienting and frustrating to break that sequence. The usual culprits here are registration processes and eligibility checks. If you can't remove them, defer them; if you can't defer them, minimize them. Find out what sequence of tasks users expect and do everything you can to stick to that pattern.

If your audience breaks down into totally separate groups who do completely different things on your website (like "doctors" and "patients"), this can be a useful first step.

The problem is that many audiences have similar or overlapping tasks. If your company provides information for journalists on its website, you'll need to give them company background information, press releases, new product information, press photographs, annual reports, and staff biographies. A financial analyst wants almost the same information. If you don't have unique audiences, you probably shouldn't label by audience.

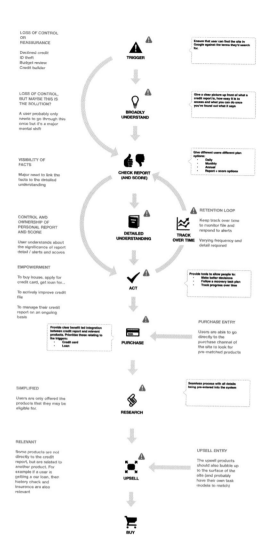

LOSS OF CONTROL OR REASSURANCE

Declined credit
ID theft
Budget review
Credit builder

TRIGGER

Ensure that user can find the site in Google against the terms they'd search for.

LOSS OF CONTROL, BUT MAYBE THIS IS THE SOLUTION?

A user probably only needs to go through this once but it's a major mental shift

BROADLY UNDERSTAND

Give a clear picture up front of what a credit report is, how easy it is to access and what you can do once you've found out what it says

VISIBILITY OF FACTS

Major need to link the facts to the detailed understanding

CHECK REPORT (AND SCORE)

Give different users different plan options:
- Daily
- Monthly
- Annual
- Report + score options

CONTROL AND OWNERSHIP OF PERSONAL REPORT AND SCORE

User understands about the significance of report detail / alerts and scores

DETAILED UNDERSTANDING **TRACK OVER TIME**

RETENTION LOOP

Keep track over time to monitor file and respond to alerts

Varying frequency and detail required

EMPOWERMENT

To buy house, apply for credit card, get loan for...

To actively improve credit file

To manage their credit report on an ongoing basis

ACT

Provide tools to allow people to:
- Make better decisions
- Follow a recovery task plan
- Track progress over time

Provide clear benefit led integration between credit report and relevant products. Prioritise those relating to the triggers:
- Credit card
- Loan

PURCHASE

PURCHASE ENTRY

Users are able to go directly to the purchase channel of the site to look for pre-matched products

SIMPLIFIED

Users are only offered the products that they may be eligible for.

RESEARCH

Seamless process with all details being pre-entered into the system

RELEVANT

Some products are not directly to the credit report, but are related to another product. For example if a user is getting a car loan, then history check and insurance are also relevant

UPSELL

UPSELL ENTRY

The upsell products should also bubble up to the surface of the site (and probably have their own task models to match)

BUY

Mapping users' behavior will help you see how to organize your software.

Hard edges

When you need to organize a group of equivalent items (like books in an online store), choose clear categories with labels that make sense to your audience.

When I first started working on Peugeot's website, information about each car was organized into features (fitted as standard), options (fitted by the dealer), and accessories (fit yourself).

This made perfect sense to the company, but when I asked them to sort a CD player, electric side mirrors, and an automatic gearbox into those categories, they couldn't agree.

The features, options, and accessories categories indicated whether something was standard—something only an insider could possibly know. If you organize items by a quality, you'll often run into these kinds of problems because users make different choices depending on their point of view.

Another way to organize the information would have been to sort it by type, such as comfort, technology, and storage. But defining these categories also depended on the user's point of view. For some customers, climate control was technology; for others it was comfort.

Simple organizational schemes have clear boundaries—hard edges— so that users know exactly where to find what they're looking for. Ask a handful of users to sort items into the categories. If they come up with different answers or if they can't easily decide, you're in trouble.

Because cars are physical objects, I decided to use the layout of the car to organize the information: interior, exterior, and performance. All the customers knew where the CD player, the side mirrors, or the automatic gearbox belonged.

Sometimes, you'll come across something that you have to place in two categories. Too much duplication leads to confusion, but occasionally it is unavoidable. Tomatoes are a fruit, but they're not sweet, and you normally find them among the vegetables at a supermarket, so they must appear in both categories. The simplest categorization is usually the one with the fewest duplicates.

Features

Options

Accessories

Good categories have hard-edged distinctions.

Alphabets, popularity, and formats

There's an old joke: Where does "finish" come before "start"?
In the dictionary.

Alphabetizing a list jumbles items. So while alphabetical lists look simple, they're often hard to use. If users don't know the correct word for what they're trying to find, they're lost. Are you looking for a jacket or a sport coat? Do you want to speak to someone in Marketing or Sales and Marketing? Alphabetical lists work well for indexes of proper nouns—where there's a "correct" word to describe something—like surnames or countries.

Arranging items by popularity is also problematic since we're unlikely to know what other people like. If I'm looking for pasta in an online supermarket, then I expect to find capellini, linguini, and spaghetti near each other (because they're all long and thin). And I'd expect to find an 18oz packet of spaghetti next to a 9oz packet—even if the 18oz packet is twice as popular. A list of popular pasta types would feel as random as an alphabetical list. Look at the shelves in a supermarket, and you won't see a label saying "long, thin pasta," but that's how it's organized. Uncovering those "hidden" categories of metadata is important in creating simple experiences.

Arranging content by format (words, pictures, videos) is another way of categorizing that looks simple but turns out to be unhelpful in the real world. If you're reading about Hawaii, you want to see photos and videos then and there. Going back to the start to find videos is just too much work.

There are a few situations I've come across where organizing by format makes sense. Examples are conference programs in which some formats, like tutorials, require a different registration process. In other words, some formats were *used* differently by the participants. But these are exceptions—it's usually simplest to organize conference information by time.

What matters is arranging items according to *relevant* indexes. Alphabets make sense for staff names, popularity makes sense for top-100 movies, format makes sense for conferences. Choose indexes relevant to the task.

Alphabetizing often jumbles items.

Patterns and anchoring

Find the right way of organizing information, and you can greatly simplify the user's experience. The right pattern helps users recognize information at a glance and enables you to strip away all but the essentials.

Take a quick look at the numbers on the next page. Memorizing the first set of numbers seems like a hard task because there's no pattern or structure to them. Once you realize that they're just the numbers from zero to nine, then memorizing them is no trouble at all.

If you can organize information around a relevant pattern that's already familiar, then users can recognize and process them in a flash. Finding those patterns is a powerful key to simplicity.

Some patterns, like numerical sequence, are so widely understood that users can't help but recognize them. Other patterns are harder to identify, especially when the user can imagine several plausible explanations. Users tend to try to see patterns, even where there are none, and it's surprising how far their imagination will carry them—like people seeing a face in a slice of toast. You can guide people to the right answer by "anchoring" them—setting their expectations with a keyword or image. A seemingly random list, like "screwdriver, telephone, scarf, master," makes sense (to those who know) if you anchor it with "Doctor Who." Anchoring users' attention and expectations with words, images, or sounds is a great way of making something feel familiar and simple.

3 **8** ⁶ **2**¹ 5 **9** 4**7**

1 2 3 4 5 6 7 8 9

**There's a simple
pattern behind this
sequence of numbers.**

Search

There are a couple big myths surrounding search and simplicity.

The first is that some users find searching easier than browsing—that there's a subgroup of people who always prefer to search. It's one of those myths that feels right. However, when Jared Spool tested a group of 30 users in over 120 shopping tasks, he didn't find a single individual who always preferred to search.

Instead, he found that when websites didn't offer links that looked like a good bet, users would search. That's not so surprising when you consider how much effort it is to think of an exact search term, type it in, and pick out a helpful search result. It's much easier to click a link that looks like it'll carry you in the right direction. Browsing requires less mental effort up front; people will usually take the path that avoids having to think too much.

The exception is when you're asking people to pick one known item from a very large number of similar items, such as a specific track from the millions of downloadable tracks on iTunes. In this situation, yes, people will tend to search. In that case, browsing is more daunting than searching.

One of the hidden benefits of browsing is that when people look over the main links on a website, or the controls on an interface, they get a sense of scope. Who needs introductory help messages when the interface speaks for itself?

The other myth is that designing a search is easier than organizing links to content. Perhaps it's because sites like Google make searching look effortless that we assume it is easy to do. My experience is that it's harder to create a simple search interface. You need to take into account spelling mistakes and synonyms in users' search terms. Also, the search results themselves need to be organized. Take a look at a Google results page and you'll see a sophisticated layout that has been chosen to match the contents of your search.

If you're designing a simple user experience, it's usually best to begin with the basic organization and then move on to designing search.

The search feature is often harder to use and harder to design than the browse feature.

Time and space

Setting events on a timeline is simple and powerful. It works best if the events are of similar duration so that users don't find themselves zooming in and out of a timeline or calendar very often. Although there may be other ways to organize the same content (such as conference themes), organizing events by time gives your audience a clear way to make sense of things.

Physical objects like hotels and countries can all be organized by space, as long as the users are familiar with the layout. For instance, you can organize a hotel website by an imaginary walkthrough of the hotel: concierge, front desk, dining, meetings and events, gym, rooms, suites. People have reasonably good memory for spaces, so this is often a good choice.

Visualizing time and space in diagrams can create some problems.

If you're plotting company offices or holiday destinations, you have to cope with the fact that some areas, like Europe, will be very crowded, while others, like the Pacific Ocean, will be almost empty. The same goes for plotting events on a day planner (not much happens between 1 a.m. and 5 a.m.).

Sometimes it's useful to see variations in density, such as seeing that there's a concentration of bus services around rush hour. Other times, it can make information hard to pick out. I can set my computer's clock by clicking a map of the world—but Paris and London are just a few pixels apart, even though they're in different time zones.

Timelines are a universal way of organizing events.

Grids

It's remarkable how far a tidy layout can go in making a design feel simple.

The form on the opposite page (top) is an interface for searching for train tickets that my company designed. It worked fine in user testing, but people hesitated over it. We revisited the layout and decided we could simplify it. We looked at the number of imaginary horizontal gridlines used to line up the field and simplified them. We also got rid of the heavy blocks of color that marked out the areas of the field and let the white space and alignment to the imaginary grid do the job.

The result was a layout that felt simpler to use, even though we hadn't changed the labels or programming of the form at all.

Lining up items using an invisible grid like this is an effective way of drawing the user's attention across the screen. It says, "Here's where to look next," without relying on bright colors or flashing images. The simpler the grid, the more powerful the effect.

Having even a few elements out of position can spoil a grid. In the example opposite, only three of the seventeen controls were out of position, but this was enough to disrupt the layout.

Grid layouts can feel regimented and constricting. One way around this is to make the layout asymmetric—for instance, by having an odd number of columns. Another is to have a few elements that straddle several columns. Take a look at websites and magazines like *Wired* or *The Guardian* online, and you'll see they're really designed around a regular, asymmetric grid.

Before

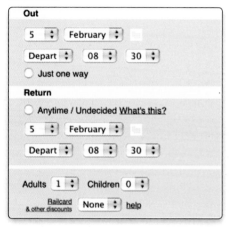

After

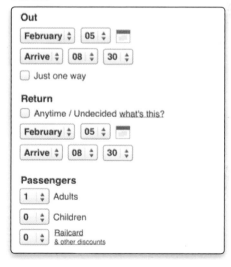

Size and location

When you're laying out items on your grid, here are some tips for sizing and positioning.

Make important things big, even if that means making them out of scale. The illustration opposite is similar to one featured in one of the first books on interface design I read—Apple's *HyperCard Stack Design Guidelines.* If you're designing a sports news website, then making the golf ball as large as the soccer ball may not be accurate, but the alternative would be to make it look as though the Masters was less important than MLS. Sports fans can debate that, but sports editors would prefer to give them equal prominence.

Less important items should be smaller. Emphasize the difference in importance as much as you can; otherwise, the user will get distracted. A good rule is: If something is half as important, make it a quarter as big.

Put similar things close together. This may sound obvious, but the benefits are huge. By placing similar items near each other, you reduce the need for visual clutter (such as color coding, labels, or boundary boxes) to explain how they are related. You also make it easier for users to focus their attention, because they don't have to look all over the screen.

When it comes to laying out navigation on computer screens, I've never seen any real evidence that navigation bars work better across the left or right side of the screen or across the top—certainly not for websites. What really matters is that users can easily find the buttons they want, and for websites that often means putting the important links right in the middle with the content.

However, for touch interfaces it can matter a lot. Putting an app's navigation at the bottom of the screen means users can touch it without covering up the screen with their hand. On large touch screens, putting navigation on the left or right risks causing problems for right- or left-handed people, respectively.

Distorting the
ball sizes shows
that each sport is
equally important.

Layers

The London Tube map crams a lot of information into a very small space. Over three hundred stations on thirteen lines are squeezed onto a pocket-sized map. One way the map stops all this information from getting jumbled up is by using a technique called perceptual layering.

Each tube line has a distinct color and seems to sit on its own layer. Without noticing, readers tune in to the color of the line they're interested in and exclude other lines from their conscious thought. Although the map is a knot of different lines, the different colors allow readers to focus on just one at a time.

You can use perceptual layering to place several elements on top of each other or alongside each other; for instance, you might use a colored tint area to connect related content. Or you can tie together elements that are scattered across a user interface, making the Buy button the same color as the shopping basket icon. If you use perceptual layers, you don't have to divide an interface into strict zones.

Perceptual layers work well with colors, but the same trick can be used with shades of gray, size, or even shapes.

Tips:

- Use as few layers as possible. The more complex your content, the fewer layers you can get away with.

- Consider putting some basic elements on a general background layer, because it can be difficult to put one item on two layers.

- Make the difference between each layer as great as possible. Readers will struggle to tell the difference between 20 percent gray and 30 percent gray. Likewise, think of color-blind users when you're choosing colors.

- For categories that are more important than others, use bright, saturated colors to make them pop off the page.

- For categories that are equally important, use perceptual layers with the same brightness and size but vary the hue (like the lines on the London Tube map).

A quick way to figure out if your design is working is to squint at the screen and see if the layers are distinct.

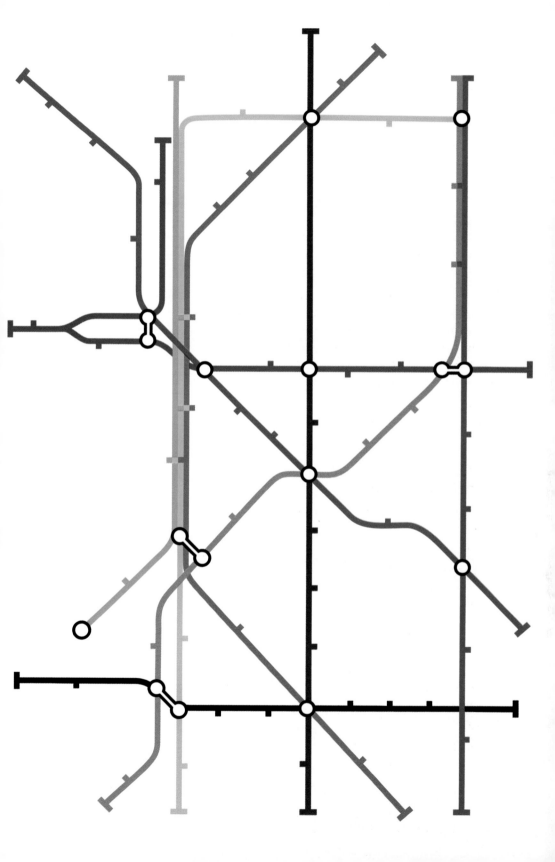

Color coding

Color coding is widespread. You see it in hospitals, folders, traffic lights, size charts, maps, dashboards—everywhere.

Perhaps because designs like the London Tube map are so successful, we tend to think color coding is a route to simplicity. But using colors to layer information is subtly different from using color to label information.

Layering information using color takes advantage of the way the mind works, so it places very little load on the user. But using colors to label information comes with a cost: Like all codes, it takes time to learn to decode, so it requires extra effort from the user.

Casual visitors may not have time to learn your code. The more colors you use, the longer it will take to learn. And if you are not rigorously consistent in using the colors throughout your design, users won't be sure what the code means.

Another problem is in taking a system that's well known in one context and using it elsewhere. For instance, some British food labels include traffic light colors to suggest whether they contain items like salt or fat that people need to limit. While the traffic-light colors are familiar to drivers, their meaning needs to be explained all over again to food shoppers, so not much is gained. And because the red and green colors don't work well for many color-blind people, they're not a universal solution (real traffic lights use position as well as color as part of their signal).

Adding color when it is not needed creates confusion.

Color coding works best when you are sure people will spend a long time learning and reusing your design or when you're using a code your audience has already learned.

You'll have to eat a lot of sushi to learn the color code by heart.

Desire paths

The next time you're in a park or a stretch of grass that's visited by a lot of people, keep an eye out for two things. First, look for the footpaths that a planner or architect has laid down through the park. These paths probably show how a designer, from an aerial view, thought people should move through the space—often in straight lines or a tidy, geometric pattern. Then look for the tracks that people have made as they wander across the grass. These well-worn "desire paths" are often quite different from the paved routes.

Looking down on his plans, the architect thought he'd designed the perfect layout. But when you walk through the park, you can see exactly why people have created the desire paths—taking a shortcut to a gate, avoiding a poorly lit corner, linking up two parallel routes. Walking the desire paths always feels simpler than sticking to the "proper" footpaths.

If you're plotting the user's path through your software, it's important not to fall in love with the neat lines and tidy organization you see in your plans.

Walk through the software repeatedly, and see what catches your eye (squint at your screen layouts!). Look through your metrics and log files. Watch other people doing the same thing. Ask "Why are you taking this path?"

Simple organization is about what feels good as you're using the software, not what looks logical in a plan.

People won't
always follow the
path you set.

Part 6

Hide

Hide

Hiding features behind a hatch or sliding panel is a popular solution to the problem of simplifying a TV remote. I have owned several remote controls that take this approach.

Another way to hide buttons is to use a touch-screen remote control. In those designs, the most frequently used features are on display, and the rarely used features are hidden in menus deeper within the device.

You can buy those kinds of programmable touch-screen remote controls—they're sold on ease of use, and they cost several hundred dollars. That shows just how far some people are prepared to go for simplicity.

Whether you go down the expensive high-tech route or add a couple of cents to the cost of your remote control by hiding features behind a plastic hatch, hiding has a big advantage over organizing: Users aren't distracted by unwanted details.

For some people, hiding is a first step to removing an unloved feature: Hide it, let it wither in the dark, and then kill it. I'm dubious of that approach. Terminating any feature means you'll need to go through the arguments I discussed in Part 4, "Remove," whether or not you've hidden it first. It's usually better to end it quickly.

Hiding anything means putting a barrier between the user and the feature, whether it's a plastic door on a remote control or a sequence of clicks on a website. You must carefully choose what to hide so as not to inconvenience the user.

Hiding some features is a low-cost solution. But which features should you hide?

Infrequent but necessary

Features that mainstreamers rarely use but that they may need to update are usually good choices to hide. These features are unlikely to appear in the story you wrote in Part 2, "Setting a Vision," as they're not related to the users' goals so much as who they are or where they are:

- Account details (for instance, setting up your server details or your signature in a desktop e-mail application)
- Options and preferences (such as changing the units in a drawing application from inches to centimeters)
- Location-specific information (such as time and date, although frequently these can be updated automatically)

Failing to include these kinds of controls in your website or application would often make it too general to suit the users' needs.

You'll see settings controls tucked away at the edges of software, away from the important stuff that tends to live at the top or the center of the screen. Settings are best located on opening pages or on all pages (it's impossible to know when users will want to change these settings, so hide them at the start of a website or at the edge of an application).

When you're looking for features to hide, settings are always a good choice. They are different from infrequently used tasks largely because tasks are focused on an external goal (such as sending a message to a friend), whereas settings are focused on using the software well (such as automatically formatting bullet lists).

**Profile settings
that change
infrequently are good
choices to hide.**

Customizing

One approach that I'm not keen on is letting users customize the interface by hiding features according to their needs. To me, this smacks of laziness and indecision on the part of the designers.

It sounds fair and generous to give the user a choice. The trouble is that customizing is time-consuming and cumbersome. You can customize the myriad floating palettes and toolbars in Microsoft Word should you have the time. But it's a laborious process that requires an understanding of what makes a good user interface. The irony is, you need to learn the vast range of features on offer before you can simplify them.

Even simple interfaces can be painful to customize. My TV receiver lets me shuffle and hide the channels I see on the program guide. This is useful, as the default order seems to be completely random. However, doing this for sixty channels takes hundreds of clicks on the remote control. It's mind-numbingly boring.

Mainstreamers do customize their devices. But they're more interested in personalizing—changing their computer desktop to a picture of their dog—than in redesigning the user interface.

Customizing is more practical when the tools for customizing are simple and when people are adding a few items, rather than rearranging many.

Google and Facebook are good examples of this. However, one of the goals of Facebook is self-expression. Choosing what content to put on your profile is part of the task. So it doesn't seem like a chore.

Customizing can also work when the user is gradually making small changes, such as adding apps and shuffling icons on a smartphone. Even then users often complain that things get out of control as time goes by and they add too much stuff.

In general, users shouldn't have to customize their software. The goal of a word processor is writing. Sifting through features and deciding which ones to show and hide is for experts.

Allowing users to customize their user interface assumes that they will be able to create effective, efficient layouts.

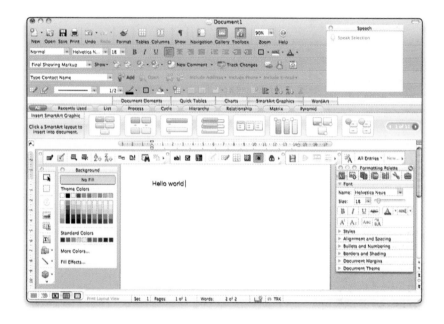

Users can customize
all these buttons in
Microsoft Word. But
is that how they want
to spend their time?

Automatic customization

Some programs try to show and hide features automatically depending on the users' behavior.

The "adaptive menus" in Microsoft Office 2000 show how rocky that road can be.

The idea was that the top-level menus would display only a subset of commonly used commands. If you left your pointer over a menu for a few seconds or if you clicked on a chevron at the bottom of a menu, it would expand to show you the full set of commands.

As you used the menus, they would remember which commands you used most frequently and adapt so that your favorite commands were visible and the others were hidden.

I recall that a few days or hours after someone had installed Office 2000, he would start walking from desk to desk asking how to turn off this feature (it wasn't easy). Microsoft dropped the idea a few years later. The BBC abandoned a similar attempt to auto-customize its home page.

Instead of making an interface simpler, automatic customization can make it more complex and frustrating to use for three reasons:

- It's hard to get the default menus right. Although most people use only a fraction of the functionality of a large application like Microsoft Word, the features they use vary widely. So what's right for one person is wrong for most others.

- Short menus make users look twice for each feature—first on the short menu and then again on the long menu. A delay or an extra click to bring up the long menu increases users' irritation.

- Users can't learn where to find items because the items' positions keeps changing.

Unless the algorithm you're using is perfect (and nothing is perfect), it will be wrong often enough to undermine users' confidence and make your interface feel complex and confusing.

Imagine what it would be like if someone rearranged your closets every night while you were asleep. That's how annoying automatic customization can be.

Default

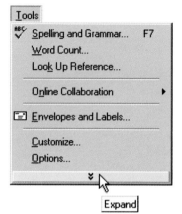

Expanded

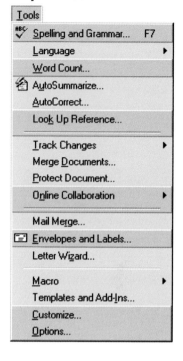

Progressive disclosure

Often a feature has a few core controls for mainstreamers and extended, precision controls for experts. Hiding the precision controls is a good way to keep things simple.

The Save dialog box is a classic example of this. The basic feature is nothing more than two core questions:

- What would you like to call this file?
- Where, from a list of options, would you like to save it?

But experts want something richer: extended options to create a new folder for the document, to search their hard disk for places to save the document, to browse their hard disk in other ways, and to save the file in a special format.

Rather than showing everything, the Save dialog box opens with the mainstream version but lets users expand it to see the expert version.

The box remembers which version you prefer and uses that in the future. This is better than automatic customization because it's the user who chooses how the interface should look.

This is also better than regular customizing because the user makes the choices as she goes, rather than having a separate task of creating the menu. This means mainstreamers aren't forced to customize.

That model, of core features and extended features, is a classic way to provide simplicity as well as power. For instance, mainstream computer users know to left-click on interfaces to make something happen. Experts know that right-clicking will bring up a menu of additional options.

Google's Advanced Search features include keyword search, in-site search, Boolean search, language-restricted search, region-specific search, linked pages, file-type restricted search, date constraints, copyright constraints, prioritization of keywords, "safe" search, and comparison search. Of these, only keyword search is visible on the main interface; the others are hidden. User testing will tell you if you're getting it right or not.

Set users' expectations with clear cues in the right context.

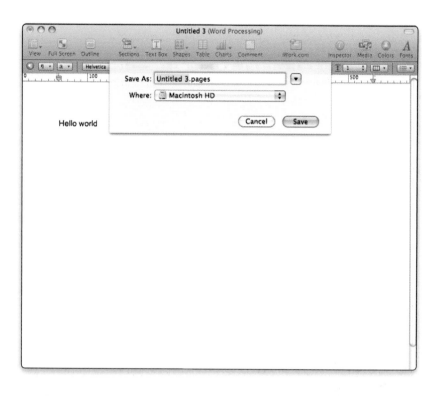

Staged disclosure

One alternative to hiding features in another part of your software is to reveal features as the user progresses deeper into the interface.

This approach works well when all users naturally seek more complexity as they progress. For instance, a user might start searching using a simple text box and then see filtering and sorting options on the results screen.

Staged disclosure is often required for processes such as booking forms, but there are rules:

■ Set the scene. When we tested one online checkout process, users found the transition from shopping cart to checkout disorienting. When the same process started with the words "Welcome to secure checkout," their problems vanished.

■ Tell a story. Users expect the sequence to unfold like a story, find out what that story is, and follow it. One online order form I tested started by asking users to enter their name and address. The owner explained that if there was a problem at a later stage, the company would still be able to contact the customer. But customers hated it. When the sequence followed a simple story ("What do you want? Now where should I send it?"), the conversion rate increased.

■ Speak the users' language. Processes tend to exist because the user has to conform to a bureaucratic process (like a passport application) or a technical procedure (like setting up a modem), and bureaucracy and technology breed jargon. For insiders, jargon is compact and specific. For novices, one unfamiliar word of jargon is more complex than an entire familiar sentence.

■ Reveal information in bite-sized chunks. If the chunks are too big, users feel the form is too complex. If the form is divided into lots of tiny nibbles, users feel the form is inefficient and tedious. Each chunk should be complete and self-contained (for instance, don't divide the address across two screens).

Wizards are a form of staged disclosure, but they often break all these rules: They fail to tell a story, use jargon without explanation, fail to explain consequences, fail to set the scene, and divide the task into chunks that are too big or too small.

Staged disclosure works best when the steps in the process meet users' expectations.

If you get it wrong,
users feel like they're
lost in a maze.

X doesn't mark the spot

A few years ago, I was reading the *New York Times* online and I came across this passage:

> This is a week of suspended animation in the city, in between holidays, when the great systems of New York—the schools, the courts, the communications media, Wall Street, City Hall, the bodegas in Queens—slow to an administrative crawl or shut down altogether.

Being English, my first response to this was, "What's a bodega?" So I did what I normally do when I don't know a word: I highlighted it so I could copy and paste it into Google. Something surprising happened: A little question mark icon appeared next to my highlight. I clicked it and up popped a window with a definition of the word (bodega: A small grocery store, sometimes combined with a wine shop, in certain Hispanic communities). The same thing works for every single word in the article; it doesn't matter if you click on "media," "animation," "the," or "a"; there's a definition for every word.

What's clever about this is that the feature is hidden, but it reveals itself precisely when you need it.

It takes courage to hide things as completely as this. The design team must have worried that users would never find their work: "We've gone to all this trouble, we should *show* people this feature."

The thing is, overemphasizing hidden features can lead to a mess. Think what would have happened if the *New York Times* had not hidden the dictionary so well.

If they had put hyperlinks in the text, they would have distracted and irritated readers. If they had put hyperlinks on every word, the page would have been a complete mess. If they had picked out a few words, they would have faced the expensive task of editing every article to decide which words were worth defining. By trying to show off the feature, they would have been dragged into a messy, ugly, expensive quagmire.

Sadly, most attempts to hide features are like this. It's like hiding buttons on the TV remote control behind a glass door.

The *New York Times'* solution illustrates what you must do to hide successfully. First, hide something as completely as possible. Second, make features reveal themselves just when they're needed.

This is a week of suspended animation in the city, in between holidays, when the great systems of New York — the schools, the courts, the communications m[?]a, Wall Street, City Hall, the bodegas in Queens — slow to an administrative crawl or shut down altogether.

The *New York Times'* dictionary feature is hidden until you highlight a word.

Cues and clues

Choosing a label for hidden features can seem tricky. How do you explain the complex and subtle extras that have been hidden?

Often you'll find extras hiding behind a vague label like "more" or a patronizing one like Google's Advanced Search. Although this is a common solution, it's not ideal. One of the reasons for hiding complexity is to prevent the user from feeling stupid. Labeling a button "advanced" effectively tells the user that she's not qualified to go there. That's not a feel-good moment.

An alternative is to use a label that will only appeal to a certain group. If you look at most computer manufacturers' websites, you'll see they swamp customers with technical details. I've watched mainstreamers lose confidence as they read about L2 cache and motherboard speed options, but to experts this is essential information.

Apple's website presents its products in a breezy, magazine style that suits mainstreamers. But in one corner is the label Tech Specs. Mainstreamers stick to the pictures and headlines. But for customers who really want to know about graphics processors, this link pops off the page. It's a phrase that they're attuned to but that mainstreamers aren't interested in.

Adobe Illustrator has a subtler solution. Some of the drawing tools have advanced features that are indicated by a small arrow on the tool palette. You click once to select the basic tool, or you click and hold to see the advanced options.

What's nice about this approach is that it is an invitation to explore, rather than a label that attempts to explain what comes next. It's also specific: The context sets the expectation that the additional functionality has something to do with the nearby tool. Experts are happy to follow those invitations because they like to explore and learn. Mainstreamers are happy to put off exploring until they have gained confidence or until they need to. No one is labeled as inadequate.

Interfaces that hide well are elegant; they use the most subtle cue possible to suggest the location and nature of the additional functionality.

**Small cues can hint
at hidden features.**

Making things easy to find

Where you place a label is more important than how big it is.

Keith Lang, who designed the interface for Skitch and Comic Life, an award-winning drawing application, points to two examples from his own work. "We gave Skitch advanced features you could get by holding down a key on your computer," he says. "We wanted to reveal these features, so we put a pop-up help box at the top of the screen whenever you used the toolbar, but most users don't see it even though it's pretty big. On Comic Life we used a small label on one of the tool palettes to explain how to use it, and that works well."

The difference is down to what Jef Raskin (one of the original creators of the Macintosh) calls the user's *locus of attention*—the area of the screen on which the user is concentrating.

When a user first looks at a screen or begins a new task, her locus of attention is wide. If you watch eye-tracking studies, you will see users scanning all over a screen when they encounter a new website. As the user concentrates on a task, her locus of attention narrows. In eye-tracking studies, you'll see users focus in on one or two areas of the screen or start reading along a body of text after they've made their initial assessment. When users have a problem, they tend to concentrate even more on the problem area of the screen. (In *The Humane Interface*, Raskin cites this as the reason that users often can't find help when they need it: They're too busy concentrating on the area of the screen where they're having a problem.)

What Keith Lang found with Skitch was that the large label placed far away from the locus of attention was ineffective. In Comic Life he found that a small label placed within the locus of attention worked well.

In the *New York Times* example, the question mark icon appeared directly over the word that I'd selected, in the center of my locus of attention. Even though the hidden feature was revealed unexpectedly, it was unmissable.

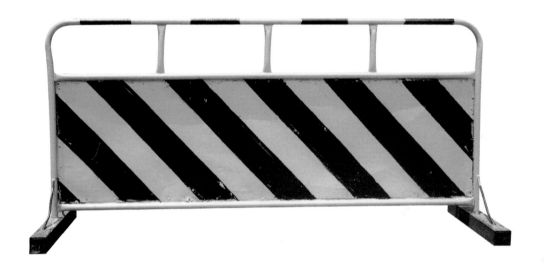

**Make sure your users
will encounter a clue
on their journey. But
don't block their path.**

After you hide

Hiding, then, depends on four things:

- Hide one-time settings and options.

- Hide precision controls, but let expert users choose to keep them revealed.

- Don't force or expect mainstreamers to customize, but offer this option for experts.

- Hide elegantly; that is, hide completely and reveal just in time.

The three strategies so far—remove, organize, and hide—fit together neatly: Remove what you don't need, organize what you do need, hide what you can. But the final strategy, displace, is really about rethinking the interface entirely.

**Hiding works,
as long as no one has
to seek too long.**

Part 7

Displace

Displace

The fourth strategy for simplifying the TV remote control is to cheat.

Designers who take this approach strip the remote control down to a few basic actions, like play and pause, and manage all the other features via a menu on the TV screen. The remote control itself is approachable, easy to understand, and simple to use.

One advantage of this strategy is that it makes good use of the remote control. Users have only a few buttons to learn, and they can easily be distinguished by touch—so it's easy to use in the dark while you're watching TV.

It's also far cheaper to make use of the existing TV screen than to add an expensive display to the remote control. The TV screen is well suited to this. It can display an infinite number of different menus, and it's bound to be in a location where the user can see it clearly.

The disadvantage of this approach is that if you displace all the features, then it's hard to guess what the remote control can do. If you had to find and access the volume controls by navigating into a menu, that might seem obscure and tedious. That's why most people end up leaving a few basic controls on the remote.

Also, though you've simplified the remote control, you now have the problem of designing a simple onscreen menu system (using the strategies of remove, organize, and hide).

But if you understand the trade-offs, displacing the right roles to the right devices works well. One of the secrets of creating simple experiences is putting the right functionality on the right platform or part of the system.

Why not take some buttons off the remote and use an onscreen menu instead?

Displacing between devices

What's easy on one platform can feel complex on another.

RunKeeper is a smartwatch app, smartphone app, and website, which together let users track their exercise routines. On your smartwatch you just tap Go, and it estimates your distance (based on the number of steps taken), split times, heart rate, and estimated calories burned. If your smartwatch has GPS, then it records your route. On a smartphone, you press Start on the app and take the phone with you. It's rather bulky to carry, but it does have GPS, so you get extras like the profile of hills run, but no heart-rate readings. Still, in both cases, capturing that data is as easy as pressing a button.

If you want to enter data on the RunKeeper website, you can enter most (but not all) of the above using an online form. It's a slow process of drawing a path on a map and filling in numerous fields.

When it comes to looking back at your data or sharing it with friends, it's easiest to review the detailed results sitting at your desk using a laptop or a big screen. You can share with friends and find the detailed results on the smartphone, but they're buried more deeply. Your smartwatch carries only a brief summary of data and no way to share with friends.

RunKeeper is a single service, but it is different depending on the device you're using; it displaces activities to the devices where they seem most appropriate. The smartwatch is small and has specialized sensors that the smartphone and desktop don't have, but its tiny screen is awkward for reviewing data or complex interactions—things that are better done on the larger devices.

The best device is the one you have with you, and at the end of a run that may be only the smartwatch. Being denied access to some parts of the service can feel awkward. So, ideally, you'd want to be able to access the entire service on the watch. Perhaps by the time you read this, you'll be able to speak to your smartwatch and ask it detailed questions. But today there's a trade-off to be made. By constraining the service on the smartwatch, the interaction is simple, and simplicity of access comes before completeness of service.

Although the service has different capabilities on each device, the overall service feels simple. The secret is making the important features easy to access and being prepared to displace the features that would overcomplicate things.

RunKeeper is a single service expressed differently on different devices.

Desktop vs. mobile vs. wearable

Some of the limitations of different types of device will change as technology gradually improves. But devices will always have strengths and weaknesses. Some activities just seem to have more "friction" when you try to complete them on the "wrong" device.

Sometimes it makes sense to displace some parts of a task, like entering data, onto a different platform. However, the best device is always the one that's closest to hand. People will put up with a certain amount of friction if it saves them the bother of finding or opening a different device. People have written entire books on their smartphones because they've been able to use them on their commute, even though the typing experience isn't the best.

If you're designing for mobile and desktop, then your best bet is to create something that works on both devices. And because mobile devices are more common and more convenient than laptop computers, design for mobile first.

Desktop/laptop	Mobile	Wearable
Photograph the user (via a webcam)	Photograph anything	May not have a camera
Input large amounts of typed text	Input small amounts of typed text	Input a few taps
Display large amounts of information on one screen	Display small amounts of information on one screen	Display glanceable amounts of information on one screen
Store very large amounts of data locally	Store moderate amounts of data locally	Least amount of local data storage
Used when seated	Used anywhere	Used "subconsciously"
Some awareness of location	Aware of precise location and orientation	May not have location awareness

Design for the strengths
and weaknesses
of each device in
your users' world.

Designing for multiple devices

The secret to fitting a service together across multiple devices is to break it down into components and check those fit with the users' needs at each step. Start by separating it into layers: the user's needs, functionality, content, and device.

Imagine you're designing a service to buy a car. Begin by mapping the stages in your users' journey: Choose a car, specify it, get finance, and so on. Don't worry about the exact order; the most important thing is to identify the stages and their approximate sequence.

Next, look at the steps users take to complete each of those stages and list them under each stage. People might get advice from friends, get expert reviews, or notice a car they like.

Now the tricky bit: Cut out references to technologies. When someone emails a friend for advice, it's not the email that matters, but their underlying need to get input from a friend. Your diagram should map users' needs at every point in the process of buying a car.

Not everyone goes through every step. Some like to ask their friends' advice; others ask professional advice. When you talk to users, you'll notice that different types of people (*personas*) fall into patterns of behavior. 'Experts' and 'mainstreamers' are good examples of personas.

Highlight the activities that matter most to the personas that represent your target audience. Now you have a list of requirements in columns: what users do, why they're trying to do it, and which users do it.

Below each column you can describe the features and content you need (or have) to support the users. You can see if you have gaps or duplications and decide what to add or remove. Over time, your functionality and content will evolve to better fit the users' needs. Draw a line below your requirements.

Underneath that, you can list the devices you intend to support (for instance, a website, a mobile app, a Facebook page) and what that means for the way the service appears on each device.

Breaking things down into layers helps you see the links between who, why, what, and how. That makes it easier to design small parts of a big system. Also, you can see where content, functionality, and even individual components can be reused, making your experience feel consistent. The bibliography has further reading on this topic.

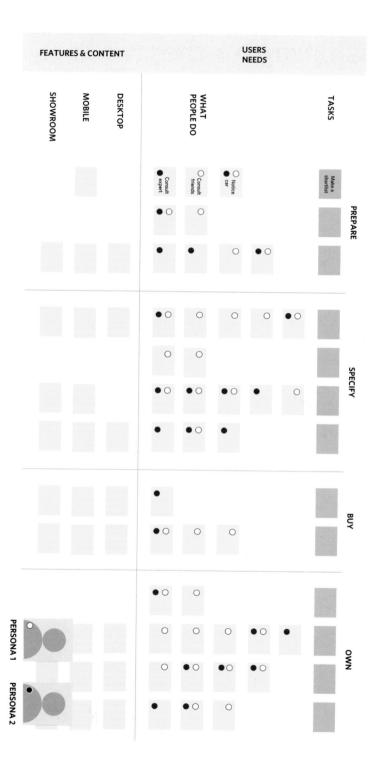

Displacing to the user

A long time ago, I was asked to design a travel planner for a tourism website. Planning a holiday is complex. Travelers have a limited amount of time, but they're often able to stretch things a little; they have a limited budget, but it will change; they are in a specific location, but they're prepared to move around; they have specific interests, but they're often looking for novel experiences. In other words, everything is up for grabs.

I decided that traveling was about managing time and space, so I began my smart travel planner with a map. I invited users to inspect locations on the map, like the Edinburgh Castle or the Science Museum in London. Users would be able to see how much time they should allow for each location. They would then add the location to an itinerary where they could rearrange the items. The smart travel planner would allow for journey times, meals, and accommodations. This meant they'd be able to see what they could fit into each day, and they would be alerted if they had tried to cram in too much.

When I tried it out, users hated it. Even though I'd designed an open-ended task, they felt that they were too constrained and that the smart travel planner was continually judging their plans. It never got built.

A few years later, I was lucky enough to be asked to try again. I chose a stripped-down approach. I let users create folders, name them as they pleased, and put whatever they wanted into the folders.

Users came up with labels I'd expected (days of the week, locations) and some I didn't expect ("under £10," "rainy days") but which made perfect sense.

It was exciting to watch users set their own success criteria. Each user did as much planning as suited her. Some users created precise travel plans; others just created lists of ideas. From the outside, some of the travel plans looked complex, but they always made sense to the users.

The complex part of travel planning is handling the ambiguity. But the simple interface had left this task to the users. I'd displaced the complexity into the users' heads.

STRATFORD-UPON-AVON

OXFORD

BATH

LONDON

The Roman Baths
A fascinating treasure trove of ancient history
with a chance to see the baths themselves,
a computer-generated reconstruction and
artifacts including a collection of Roman curses.

Mon-Fri 0900-1830 (includes Bank Holidays)
Sat-Sun 0900-1730
Christmas: Closed

£10 Adults, £5 Children / Student / Over 65

Allow [1 hour ▼]   Add this

◀ ● ● ● ● ● ▶

My Travel Plan

LOCATION	ACTIVITY	TIME
Bath	Excelsior Hotel	N/A
Bath	The Roman Baths	0930-1130
Bath	Train to Oxford	1042-1153
Oxford	The King's Head	1230-1400
Oxford	Punting	1415-1515
Oxford	Ashmolean Museum	1530-1700
Oxford	Train to London	1722-1835

Not enough time. Remove activities
or reduce time allocations

Tuesday

Kid's things

Travel discounts

What users do best

The reason the basic travel planner felt simple is because it let the users and computers do what they're best at.

Computers are good at storing detailed information accurately. You only have to tell a computer your phone number once, and it will remember it forever. People are terrible at remembering those kinds of details. The basic travel planner gave the task of remembering to the computer.

Computers are good at calculating accurately. But travel planning begins with approximate calculation and imagining how an itinerary might unfold, both of which are better done by people. The basic travel planner left users in charge of making sense of the plans.

People like to be in control of outcomes. The smart travel planner forced users to create one kind of plan. If there were too many activities, it showed a warning message—too few, and it felt incomplete. The basic travel planner left it up to the user to decide when they had done enough planning.

The basic travel planner left users with the tasks of forming a goal and deciding how to organize their notes. These are complex for a computer, but they are tasks people are good at, so the basic travel planner felt simple. The smart travel planner set a goal and forced users to organize information in a way that didn't always suit them, so it felt complex.

People will accept recommendations from computers but only if they understand *why* the recommendation is being made, approve of the recommendation criteria, and feel the risk of error is acceptable. If the recommendation fails those tests, then people won't trust it.

One of the keys to making an experience feel simple is to understand which tasks to give to the computer and which to leave for the user:

People	Computers
Setting goals and planning	Following procedures
Approximate calculation	Accurate calculation
Recognizing information	Storing and retrieving details
Seeing patterns in small data sets	Seeing patterns in large data sets
Choosing from small lists	Sorting and filtering large lists
Estimating	Measuring
Imagining	Cross-referencing detailed information

When the user is directing and the computer guiding, the experience feels simpler.

Notifications and interruptions

Computers are perfect for monitoring unpredictable or slow-changing situations because they never get bored. Most of us are happy to let our computers check for incoming messages, watch for software updates, and keep an eye on the air conditioning because it takes a chore off our hands. The problem comes when they need to notify us of a change.

Notifications seem to pop up at the most inconvenient times—when we're in a meeting, in the middle of another task, or watching a movie. One friend of mine was giving a presentation to 500 people when her laptop decided that now was the time to ask her about updating her software.

None of this feels simple. As we plug our devices into more aspects of our lives, we're feeling constantly interrupted, distracted, and harassed.

This is the price of handing off monitoring so much of our lives to computers: constant distraction, or turning off notifications and maybe missing something important.

When you're designing notifications, don't follow this pattern. Think about what a great personal assistant would do if he or she were giving you the notification: look at the social context.

If a great personal assistant wanted to remind you to update your software, they wouldn't do so as soon as the software company sent them news of the update. Instead they'd wait until you were about to head off for lunch or leave your computer at the end of the day. They'd recognize that what was important to the software company was of passing interest to you.

Our computers have many of the abilities of a great personal assistant. They can access our calendars, see when we're working, learn, and ask senders if they want to override your preferences.

Notifications and alerts should be delivered with an appreciation of social context and a degree of humility (because the software update you worked all weekend to deliver may not be the most important thing happening in the user's life on Monday morning).

Monitoring is a task computers are good at. Knowing when someone can be interrupted is a task humans are good at. If you're bridging the two, you will need to either hand off to a human or teach your computer something about social context.

Monitoring requires a computer's mindless attention. But notifications require a sense of etiquette.

Creating open experiences

Clever designers often simplify by making one component serve several purposes. For instance, in some cars, the rear windshield heating element is also the radio antenna.

You can reduce complexity in software by designing features that can be used for several purposes. The task of choosing how to use them is displaced to users.

For example, have you noticed how many ways there are to save items on sites like Amazon or your online supermarket? You can drop an item in your shopping basket for later, take it out of your shopping basket, put it in your "Saved Items," add it to a wish list, set up lists for weddings and birthdays, or look at lists of "Favorite Items." Many sites have several or all of these features.

Each of these features has specialized functions, like the ability to publish wish lists to friends. But mostly, they all do the same thing: save an item to buy later.

Users must learn several features and then remember which one they used to save an item and how to get to it again.

This also requires a lot of effort on the store's part: maintaining code, providing help and technical support, making sure it all works, and finding a place within the website for all these features to live.

When I come across similar features like this, I look to see whether they can be combined into a general tool.

Imagine if these lists were all in one place: a set of folders within the shopping basket. You could name the folders (Wedding, Birthday, Travel Books, etc.) and choose whether to publish them to a friend. One feature could do the job of four.

It's important to prompt users with ideas for the different ways the feature might be put to use. Suggesting some ideas for naming their lists would be enough to start users thinking about what they could do with the feature.

Simplifying by combining similar features is a neat solution. The result may not be perfectly tailored to each use, but there are significant advantages. It's easier to find one feature than to pick it out from several similar features. It's easier to learn one feature than several. And it's easier to maintain just one feature.

In some cars, the rear window heating element doubles as a radio antenna, simplifying the design.

Kitchen knives and pianos

Simple interfaces are ones that make sense to experts and mainstreamers alike.

Take a really simple device like a kitchen knife. A novice can use a kitchen knife to get a "good enough" outcome without much instruction or help. An expert can use the same knife to get "precision control"—chopping quickly, carving shapes, and so on. The knife is the same, but the expert's technique turns it into an expert tool.

The same goes for a piano. A novice can pick out a tune without any training and would probably tell you that it felt pretty simple. An expert can play a sonata without much trouble. The difference is technique.

What makes these experiences feel simple is that the experts and mainstreamers are free to set their own goals. They have their own expectations of how much effort it will take to achieve them, based on their experience. Playing the piano only feels like a chore when the music is at the limits of one's training.

It's the same with open experiences like the simple travel planner: They often work well for experts and novices. Letting users define success (a complete travel plan or a list of ideas) is important. So is giving them a tool that is simple enough that they can imagine how to achieve their goals.

These interfaces don't always suit the middle-ground users, who can see what an expert can do but lack the technique to get there. That explains the appeal of kitchen gadgets for chopping onions and eggs, or electronic pianos that fill in tunes with a backing track.

Those kinds of gadgets offer assistance, but the price is clutter. Imagine a kitchen without a knife but packed with specialized chopping devices.

The trick with open interfaces is to minimize the number of "handy" gadgets for the middle ground.

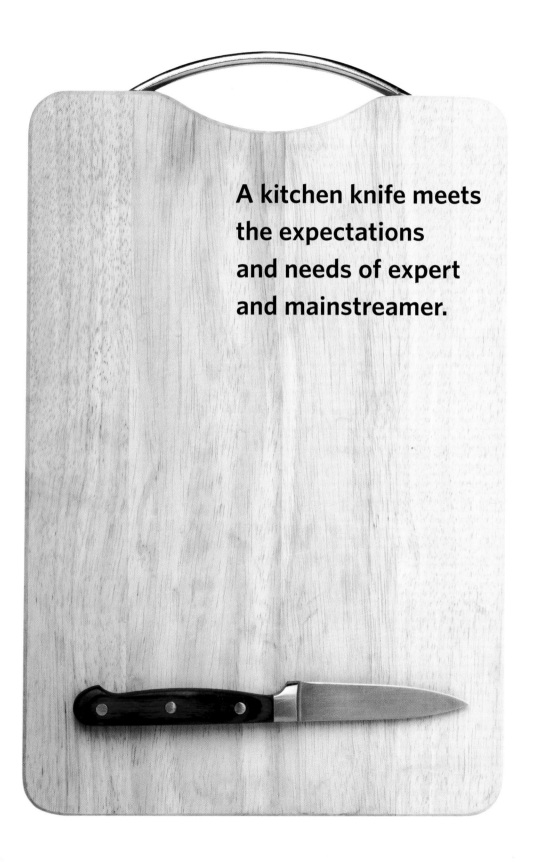

A kitchen knife meets the expectations and needs of expert and mainstreamer.

Unstructured data

Filling in forms is an irksome task that frequently feels stilted and complex. One reason for this is that users are being asked to format information so that it makes sense to the software or some distant bureaucracy.

One way of getting around this is to let the user take on the task of making sense of the data. An example of this is Ta-da List, 37signals' basic to-do list website. The creators point out that they intentionally kept data entry simple; for instance, there was no way to add a due date to a task. They figured that if people wanted they could just add "Due 17 January" to the description of the item.

This approach of letting the user make sense of his or her own notes works just fine. It's simple, open, and "human." The moral: Don't assume you need to make users fill in a structured form.

If the data needs to be processed by a computer (for instance, if the tasks need to be sorted into date order), then the data needs to be structured. But often the computer can recognize and structure the data in the users' notes.

Some email software looks for phrases like "next Tuesday" or "1-800-654-3001" in email and turns them into clickable links that create an appointment in the user's calendar or dial the number on his phone (or even automatically suggest a calendar entry).

The user is free to write emails in loosely structured, human terms. The computer takes on the task of figuring out whether there's any data in there that needs to be structured or acted on.

One of my pet peeves is online forms that require me to enter my credit card details without spaces or that reprimand me for using brackets in phone numbers. It's easy to write software that can deal with this; it's lazy and rude for companies to force customers to stick rigidly to their data-entry rules.

Let the computer take on some of the responsibility for structuring the data, and you'll simplify the user experience.

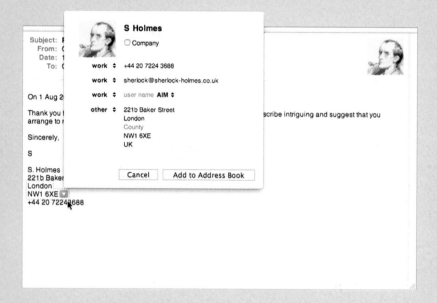

**Apple's "data detectors"
look for addresses
in emails and let you
add them to your
contact database.**

Trust

Displacing tasks is easiest when you're dividing them between two devices that must be used together in a specific way. The DVD remote control must be used with the TV display, so it's fairly easy to see what each should do.

When you're not sure how the devices will be used together, displacing becomes harder.

You can't be sure how the RunKeeper mobile app and website will be used. Some people may not have mobile phones and will just want to use the website. Some people may stick exclusively to using their mobile phone. Some people may do a bit of both.

When that uncertainty creeps in, you find yourself duplicating functionality between platforms. So it is with RunKeeper, where only a little of the functionality is displaced between website and mobile.

You need a sense of certainty to be able to displace tasks effectively.

If you're going to displace tasks to be the responsibility of the user, you must trust that the user will take on those tasks.

Trusting the audience is hard. Designers are used to watching them fail in user tests. Programmers are used to thinking of all the ways a system could go wrong so that they can design for error. Product managers want to provide users with interactive tools that take on all the hard work. And sometimes the unspoken purpose of software is to make users behave in ways that are convenient for the designer.

In other words, we often treat users like children. But in protecting users from making errors or finding their own solutions, we often deny them the chance to make their own decisions. No wonder users often feel rebellious or resentful toward computers.

The only way to build that trust is to try out prototypes and mock-ups with users. When you get the balance of tasks right—letting users focus on choosing and directing and having the computer focus on remembering and calculating—you'll create experiences that are simple and surprising because of the creativity users can bring to them.

Computers often make users uncomfortable because they control and direct users' behavior. Simple experiences require trust.

Part 8

Before we go

Conservation of complexity

Making things simple can sometimes feel like a game of Whac-A-Mole. Bash one complexity on the head and another pops up elsewhere.

Take the example of the online bank statement in Part 4, "Remove." The original design had the user choose a month and a year and then asked the bank for the corresponding statement. This was straightforward to program but felt harder to use thanks, in part, to the high risk of errors.

In the revised design, the user picked from a list of available statements. This felt simple to use and eliminated errors but was more complex to program and put more load on the bank's server as it checked which statements were available.

While he was part of the team developing the Macintosh, Larry Tesler summarized this in his Law of Conservation of Complexity:

> Every application must have an inherent amount of irreducible complexity. The only question is, Who will have to deal with it?

Designing simple user experiences often turns out not to be about "How can I make this simpler?" but rather "Where should I move the complexity?"

- Should a task be automated (like a thermostat) or controlled by the user (like a light switch)?

- Should an interface have many specific controls (like a word processor) or a few general controls (like a note-taking app)?

- Should a task be completed at one time (like signing up for a driver's license) or spread over time (like creating a LinkedIn profile)?

- Should a task be handled by the conscious mind (like using onscreen controls to filter search results) or the unconscious brain (like focusing on the green District line on the London Underground map)?

The secret to creating a simple user experience is to shift complexity into the right place so that each moment feels simple.

Bash away—
but complexity will
keep popping up.

Let the user be the star

If you overload any machine, it starts to creak and slow down under the stress. The same is true for people. Give us too many things to remember, and we'll forget; too many tasks to juggle, and we'll drop one; too many decisions to make, and we'll freeze.

Making software *usable* means not exceeding your users' capacity. However, users always demand more detailed information, more choices, more stuff—it's human nature. So the tendency is to design usable experiences right up to that maximum—stopping just short of overloading the user.

Simple experiences don't do that. They leave the user with plenty to spare. What happens to all that "unused" capacity?

A travel firm once asked my company to compare customers' experiences of researching holidays online and through brochures. Watching customers plow through websites, reviewing detailed information and options, we were struck by how tense and irritable they were.

The brochures were far simpler: some large photos of a resort, a few icons highlighting key features, and a table of prices. When they looked through the brochures, customers were relaxed, imagining what the holiday could be like. They enjoyed themselves.

Simple experiences leave users with enough capacity to think about how what they're doing fits in with their lives.

A simple password system (like fingerprint or facial recognition) lets users concentrate on picking up their phone and capturing the moment; a simple TV remote control allows users to focus on the movie.

Don't try to fill the user's mind with your design. Designing for simplicity leaves your user enough room to fill in details from his or her life and create a richer, more meaningful experience.

**Your design is the
blank page on which
the user writes.**

Bringing people with you

Simplicity requires single-mindedness, which means it requires agreement across an organization. When I talk to people at every level of design, in every type of organization, the question I'm asked most often is, "How can I convince my company to listen to me?"

I don't like the word "convince." It implies that you can magically bend someone's will toward you. In my experience, the harder you try to "convince" someone, the more barriers they put up. Even if they appear to go along with you, they'll change their mind at the first sign of trouble.

You can't convince people. They have to see for themselves.

Begin by listening to them to understand what matters to them. Respectfully summarize back to them so they know you've understood. When someone feels they've been listened to, they're far more open to new ideas.

Next, find out what information they're acting on. Chances are, there are gaps in their knowledge (or yours). Most people in an organization are distant from the customers or users. That makes it easy for them to pretend that their self-interest and users' needs are identical—that users won't mind being made to do a bit of extra work or that the problems behind poor performance are mysterious.

In most cases, the best approach is to say, "I want what's best for customers, too; let's watch some and see if it works." When they discover it doesn't work, be there for them. Give them time to process the new information. Let them own the problem (don't let them dodge it). Ask to help them to solve the problem.

You don't need permission to run user research. If your product is secret, then test a competitor's solution ("Wow. Ours has that same feature; we need to fix it"). If the public isn't allowed on-site, go to them instead. If the product manager is afraid user research will expose her bad decisions, then get her to buy into the technique by testing something less controversial first. If the head of sales won't let you talk to ten of his most important clients, then arrange for him to watch a user test with *one* of his clients who switched to a competitor's product.

Don't try to convince people. Listen to them. Let them see for themselves. Let them own the problem.

Try to persuade people, and you're likely to meet a brick wall.

Simplicity is a profound strategy

Once you start designing for simplicity, you set off a chain of events that goes far beyond design.

Rust, Thompson, and Hamilton's research (Part 4, "When features don't matter") shows that people are more likely to value the simplicity of products with fewer features once they've used them. That makes perfect sense: Simplicity is an experience.

This points to something profound. If you're choosing simplicity as your strategy, then you really need to sell the experience, not the product. That means you need to get people to experience the product by offering them free trials or getting it into their hands. You need to show how easy something is, rather than how powerful it is. You need to focus on generating word-of-mouth testimonials, rather than buying attention.

It also means that other aspects of the experience—like customer service, returns, servicing, and upgrades—need to be easy, too (so long as that doesn't lead to decisions that compromise the *use* of the product).

Pretty soon, you find yourself redesigning the entire organization. If you don't, you'll find that the organization will gradually erode the simplicity of your product, so you really have no choice.

That in turn means you need to create a *culture* of seeking simplicity, so that when they encounter new situations where there are no rules to follow, people instinctively do what's right.

Culture—the behavior, rituals, language, and unique objects of an organization—emerges from what an organization values. So your organization's key metrics for success will need to be linked unambiguously to the simplicity of its products.

The decision to pursue simplicity is the start of a long journey. How much you achieve depends on how many people you want to bring with you.

**A strategy of simplicity
can lead you deeper and
further than you imagine.**

Bibliography

Forget all the rules you ever learned about graphic design: including the ones in this book by Bob Gill (Watson-Guptill Publications, 1981) is an inspiring book about how to discover simplicity. Gill shows that the key to great design is understanding and expressing the problem in the right way. It's out of print now but is still available in second-hand book stores and is worth tracking down.

About Face: The Essentials of Interaction Design by Alan Cooper, et al. (Wiley, 2014) is perhaps the best and most detailed explanation of how to create and use personas. Many people run into difficulties with personas because they aren't aware of the rigor required to get them right. This book provides that, as well as a comprehensive guide to designing for users.

Mental Models: Aligning Design Strategy with Human Behavior by Indi Young (Rosenfeld Media, 2008) will teach you how to create the kind of diagram described in Section 7, "Managing Multiple Devices." It gives detailed advice on how to do everything from conducting field research, to identifying personas from your data, to creating and using the diagrams.

Getting real: the smarter, faster, easier way to building a successful web application by 37signals (37signals, 2006) is a characteristically gutsy book and a great place to find inspiration for the hard task of removing features (or not admitting them in the first place).

Envisioning Information by Edward Tufte (Graphics Press, 1990) is an excellent guide to presenting information in images and tables (and by extension, to laying out user interfaces and simplifying by organizing). It's beautifully produced and full of wonderful examples and wisdom; but you'll get the most from it if you sit down and read each chapter in a single sitting, like a series of lectures from a brilliant tutor.

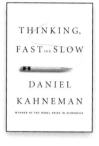

Thinking, Fast and Slow by Daniel Kahneman (Farrar, Straus, Giroux, 2011) opens a door into how people make decisions and the role that biases play in knee-jerk decision making. Understanding when and why people rely on non-rational decision making is important in designing systems that feel simple to use, and that don't mislead users.

Going Responsive by Karen McGrane, and *Responsive Design: Patterns and Principles* by Ethan Marcotte (both published by A Book Apart, 2015) are two excellent resources on creating designs that work across multiple devices. The first will help you deal with the problems you'll encounter in managing your project; the second takes a deep look at problems like managing navigation and images across multiple devices.

Designing Connected Products: UX for the Consumer Internet of Things by Claire Rowland, et al. (O'Reilly, 2015) is a comprehensive guide to designing devices connected to the Internet.

Photo Credits

Page 59, Photo courtesy Ben Stanfield

Page 61, © Roadk/Shutterstock.com

Part 3
Strategies for simplicity

Page 71, © serezniy/123rf.com

Page 73, © iodrakon/Shutterstock.com

Page 75, © Mitch Gunn/Shutterstock.com

Part 4
Remove

Page 87, © Bjoern Lotz/Shutterstock.com

Page 89, Public Domain

Page 91, © Sarawut Chamsaeng/Shutterstock.com

Page 93, © Brian A Jackson/Shutterstock.com

Page 95, © Margo Harrison/Shutterstock.com

Page 101, Public Domain

Page 103, © TRIG/Shutterstock.com

Page 105, © Sashkin/Shutterstock.com

Page 109, © eZeePics Studio/Shutterstock.com

Page 127, © karen roach/Shutterstock.com

Page 129, © Ray Yuen

Page 133, © koosen/Shutterstock.com

Part 5
Organize

Page 139, © Drozhzhina Elena/Shutterstock.com

Page 149, © Vitaly Korovin/Shutterstock.com

Page 151, © Jules Selmes/Pearson Education Ltd

Page 155, Basketball © Mike Flippo/Shutterstock.com

Page 155, Tennis ball © Imageman/Shutterstock.com

Page 155, Baseball © R. Gino Santa Maria/Shutterstock.com

Page 155, Soccer ball © Iakov Filimonov/Shutterstock.com

Page 159, Photo courtesy Adam Wilson

Page 161, Photo courtesy Andrew Skudder

Index

deleting options and preferences,
114–117

design. *See also* remote control

avoiding unnecessary text, 120–121

beginning, 22–23

chunking, 138–139

by committee, 24–25

context for, 50–51

deleting options and preferences,
114–117

refining, 74–75

removing clutter, 118–119

searching vs. browsing, 148–149

simplicity in, 212–213

smart defaults in, 112–113

triggering complexity, 96–97

designers

creating stories, 48–49

describing core user actions, 44–45

designing for mainstream, 34–35,
104–105

evaluating feature requests, 94–95

finding qualities of brand, 40–41

focusing on core experience, 88–89

getting right vision, 58–59

helping users feel in control,
36, 42–43

how to remove features, 102

ignoring expert customers, 32–33

listening to users, 4, 22, 104–105,
214–215

observing user experience, 26–27,
56, 206–207

prioritizing features, 104–105

pursuing simplicity, 216–217

reflecting on stories, 56–57

researching user experience, 92–93

sharing design vision, 60–61

uncovering user needs, 38–39

understanding work interruptions,
28–29

desire paths, 160–161

desktop/laptop device design, 190

details of design, 74–75

devices. *See also* user interface

designing for multiple, 192–193

displacing between, 188–189

distractions and use of, 28

interface for different, 190–191

"displace" strategy, 186–207

about, 80, 81, 186–187

combining similar features,
200–201

design for multiple devices,
192–193

designing for different devices,
190–191

displacing between devices,
188–189

monitoring and notifications,
198–199

passing complexity to user,
194–195

providing open interfaces, 202–203

trusting users, 206–207

understanding what people do best,
196–197

disruptive strategy, simplicity as, 4–5

distractions

affecting use of devices, 28, 29

removing clutter, 132–133

replacing with simplicity, 110–111

Don't Make Me Think (Krug), 120

E

Eames chair, 16, 17

Eames, Charles, 74

emotions

brand's emotional qualities, 40

users' emotional needs, 38–39

end-users. *See* users

errors

new features triggering, 96–97

redesigning to avoid, 98–99

ethical qualities of brand, 40
experts
    customizing for, 168–169, 182
    defined, 30
    hiding precision controls for, 172, 182
    ignoring, 32–33

F
fake simplicity, 8–9
features
    combining similar, 200–201
    cutting from products, 86–87,
        90–91, 102
    disclosing, 172–175
    displacing remote control, 186–187
    effect on user experience, 90
    evaluating requests for, 94–95
    fixing safety, 90
    hiding, 166–167, 176–177
    prioritizing, 104–105
    triggering changes in user
        experience, 96–97
    usability of product vs., 100–101
Flip camcorder, 4–5, 50, 54
Ford, Henry, 34
Form, Function & Design (Grillo), 16
formats for organization, 144

G
Google's Advanced Search, 178
Grillo, Paul Jacques, 16

H
Hamilton, Rebecca W., 100, 216
help boxes, 8, 9
"hide" strategy, 164–183
    about, 80, 81, 164–165, 182–183
    automatic customization strategies,
        170–171
    choosing features to hide, 166–167
    customizing as, 168–169, 182
    how to hide features, 176–177

labels and cues with, 178–179
making things easy to find, 180–181
progressive disclosure of details,
    172–173
staged disclosure, 174–175
Holmes, Jr., Oliver Wendell, 12, 13
home distractions, 28, 29
Human Interface, The (Raskin), 180
hyperlinks, 110

I
infrastructure wins, 70
Innovator's Dilemma, The (Chris-
    tensen), 84
Insanely Great (Levy), 58
instructions, faking simplicity with,
    8–9
iPhone, 54
iPod, 32
Iyengar, Sheena S., 108

J
jargon, 174
Jarvis, David, 90
Jobs, Steve, 54, 58, 59
Johnson, Michael, 50

K
Kaplan, Jonathan, 4
Krug, Steve, 120

L
Lang, Keith, 180
Lanham, Richard, 122
Law of Conservation of
    Complexity, 210
Lawrence, Jennifer, 46, 48, 114
layers
    perceptual, 156–157
    separating product function into,
        192–193
learning under pressure, 10–11

Lepper, Mark R., 108
Levy, Steven, 58
listening to users, 4, 22, 104–105, 214–215
locus of attention, 180
London Tube map, 156, 157

M

mainstream users
  defined, 30
  designing for, 34–35, 104–105
  displaying core controls for, 172–173
  "hide" strategies for, 182
  simplifying options and preferences, 114–117
  what they want, 36–37
maps
  London Tube, 156, 157
  mapping users' behavior, 140–141
  spatial organization using, 150
marginal gains, 74
Marriott website, 130–131
mass appeal, 34–35
Merholz, Peter, 68
Microsoft Word, 138, 168–169
Minimal Viable Product (MVP), 86
minimalism vs. simplicity, 16–17
mobile device design, 190
Model T, 34, 35
Moffett, Jack, 90
monitoring, 198–199
MVP (Minimal Viable Product), 86

N

navigation design, 154
New York Times, The, 176–177, 180
notifications, 198–199

O

office distractions, 28, 29
On Writing Well (Zinsser), 48
online sales, options preventing, 116–117
open user experiences, 200–201
open user interfaces, 202–203
Org Design for Design Orgs (Merholz), 68
"organize" strategy, 136–161
  about, 80, 81, 136–137
  alphabets, popularity, and formats for, 144–145
  categories for organizing information, 142–143
  chunking as, 138–139
  color coding information, 158–159
  finding users' paths through software, 160–161
  grids, 152–153
  mapping users' behavior, 140–141
  patterns that anchor information, 146–147
  perceptual layering, 156–157
  sizing and positioning items, 154–155
  time and space organization, 150–151
  using with remove and hide strategies, 182
outdoor distractions, 28, 29

P

patterns anchoring information, 146–147
plot, 50–51
Pogue, David, 6
popularity of products, 144
Porter, Michael, 66
positioning items, 154–155
practical qualities of brand, 40

users

adding instructions for, 8–9

anchoring attention of, 146–147

choosing searching vs. browsing, 148–149

creating enjoyment for, 212–213

effect of product errors on, 98

evaluating feature requests from, 94–95

expecting them to learn, 10–11

expert, 30, 32–33

giving sense of control, 36, 42–43, 128–129

having stakeholders meet, 24

learning perspective on simplicity, 14–15

lightening load on, 106–107

limiting choices for, 108–109

listening to, 4, 22, 104–105, 214–215

mainstream, 30, 34–35, 104–105

mapping behavior of, 140–141

memory of, 138

observing experience of, 26–27, 56, 206–207

preference for basic improvements, 88–89

pressure's effect on, 10–11, 36

reaction to complex products, 6

shortening conversations for time, 124–125

tasks best for computers and, 196–197

trusting, 206–207

types of, 30

uncovering needs of, 38–39

unstructured data from, 204–205

valuing usability over features, 100–101

what mainstreamers want, 36–37

willing adopters, 30

work interruptions of, 28–29

V

Virgin Atlantic, 40–41

vision

aligning group's, 24–25

clarifying your, 22–23

creating stories for, 48–49

distractions affecting device use, 28–29

getting right, 58–59

including in strategy, 66–67

quick way to get, 54–55

reflecting on and testing stories, 56–57

sharing design, 60–61

summing up user experience, 46–47

three types of users, 30

usability vs. simplicity, 52–53

visual clutter, 118–119

Volkswagen ad, 123

W

wearable device design, 190

willing adopters, 30

wizards, 173

work interruptions, 28–29

world, character, and plot, 50–51

Wroblewski, Luke, 58

Z

Zhu, Erping, 110

Zinsser, William, 48